Library of
Davidson College
VOID

Cryopreservation and Freeze-Drying Protocols

Methods in Molecular Biology™
John M. Walker, SERIES EDITOR

38. **Cryopreservation and Freeze-Drying Protocols**, edited by *John G. Day and Mark R. McLellan, 1995*
37. **In Vitro Transcription and Translation Protocols**, edited by *Martin J. Tymms, 1995*
36. **Peptide Analysis Protocols**, edited by *Michael W. Pennington and Ben M. Dunn, 1994*
35. **Peptide Synthesis Protocols**, edited by *Ben M. Dunn and Michael W. Pennington, 1994*
34. **Immunocytochemical Methods and Protocols**, edited by *Lorette C. Javois, 1994*
33. **In Situ Hybridization Protocols**, edited by *K. H. Andy Choo, 1994*
32. **Basic Protein and Peptide Protocols**, edited by *John M. Walker, 1994*
31. **Protocols for Gene Analysis**, edited by *Adrian J. Harwood, 1994*
30. **DNA–Protein Interactions**, edited by *G. Geoff Kneale, 1994*
29. **Chromosome Analysis Protocols**, edited by *John R. Gosden, 1994*
28. **Protocols for Nucleic Acid Analysis by Nonradioactive Probes**, edited by *Peter G. Isaac, 1994*
27. **Biomembrane Protocols:** *II. Architecture and Function*, edited by *John M. Graham and Joan A. Higgins, 1994*
26. **Protocols for Oligonucleotide Conjugates**, edited by *Sudhir Agrawal, 1994*
25. **Computer Analysis of Sequence Data:** *Part II*, edited by *Annette M. Griffin and Hugh G. Griffin, 1994*
24. **Computer Analysis of Sequence Data:** *Part I*, edited by *Annette M. Griffin and Hugh G. Griffin, 1994*
23. **DNA Sequencing Protocols**, edited by *Hugh G. Griffin and Annette M. Griffin, 1993*
22. **Optical Spectroscopy, Microscopy, and Macroscopic Techniques**, edited by *Christopher Jones, Barbara Mulloy, and Adrian H. Thomas, 1994*
21. **Protocols in Molecular Parasitology**, edited by *John E. Hyde, 1993*
20. **Protocols for Oligonucleotides and Analogs**, edited by *Sudhir Agrawal, 1993*
19. **Biomembrane Protocols:** *I. Isolation and Analysis*, edited by *John M. Graham and Joan A. Higgins, 1993*
18. **Transgenesis Techniques**, edited by *David Murphy and David A. Carter, 1993*
17. **Spectroscopic Methods and Analyses**, edited by *Christopher Jones, Barbara Mulloy, and Adrian H. Thomas, 1993*
16. **Enzymes of Molecular Biology**, edited by *Michael M. Burrell, 1993*
15. **PCR Protocols**, edited by *Bruce A. White, 1993*
14. **Glycoprotein Analysis in Biomedicine**, edited by *Elizabeth F. Hounsell, 1993*

Earlier volumes are still available. Contact Humana for details.

Methods in Molecular Biology™ • 38

Cryopreservation and Freeze-Drying Protocols

Edited by

John G. Day

The Windermere Laboratory,
Institute of Freshwater Ecology, Cumbria, UK

and

Mark R. McLellan

Acer Environmental, Cheshire, UK

Humana Press Totowa, New Jersey

© 1995 Humana Press Inc.
999 Riverview Drive, Suite 208
Totowa, New Jersey 07512

All rights reserved.

No part of this book may be reproduced, stored in a retrieval system, or transmitted in any form or by any means, electronic, mechanical, photocopying, microfilming, recording, or otherwise without written permission from the Publisher. Methods in Molecular Biology™ is a trademark of The Humana Press Inc.

All authored papers, comments, opinions, conclusions, or recommendations are those of the author(s), and do not necessarily reflect the views of the publisher.

This publication is printed on acid-free paper. ∞
ANSI Z39.48-1984 (American Standards Institute)
Permanence of Paper for Printed Library Materials.

Photocopy Authorization Policy:
Authorization to photocopy items for internal or personal use, or the internal or personal use of specific clients, is granted by Humana Press Inc., provided that the base fee of US $4.00 per copy, plus US $00.20 per page, is paid directly to the Copyright Clearance Center at 222 Rosewood Drive, Danvers, MA 01923. For those organizations that have been granted a photocopy license from the CCC, a separate system of payment has been arranged and is acceptable to Humana Press Inc. The fee code for users of the Transactional Reporting Service is: [0-89603-296-5/95 $4.00 + $00.20].

Printed in the United States of America. 10 9 8 7 6 5 4 3 2 1

Library of Congress Cataloging in Publication Data

Main entry under title:

Methods in molecular biology™.

Cryopreservation and freeze-drying protocols / edited by John G. Day and Mark R. McLellan.
     p.   cm. — (Methods in molecular biology™ ; 38)
    Includes bibliographical references and index.
    ISBN 0-89603-296-5
     1. Cryopreservation of organs, tissues, etc.—Methodology. 2. Freeze-drying—Methodology.
I. Day, John G. II. McLellan, Mark R. III. Series: Methods in molecular biology™ (Totowa, NJ) ; 38.
QH324.9.C7C77   1995   94-44732
579—dc20   CIP

Preface

The storage of biological material for regular or future use is a fundamental requirement in many biological and medical sciences. Cryopreservation and freeze-drying are the preferred techniques for achieving long-term storage, and have been applied to a diverse range of biological materials. Though the basis for many methodologies is common, laboratories frequently lack expertise with the correct storage procedures, so that many apply outdated or inappropriate protocols for storing their samples or cultures.

Cryopreservation and Freeze-Drying Protocols is a compilation of the many and varied methodologies that have been developed in expert laboratories. The protocols are reproducible, robust, and in most instances have been transferred quite successfully to other laboratories. Our intended readers are those proposing to establish or improve biostorage systems in their own laboratories or units, whether concerned with culture collections, animal husbandry, aquaculture, or human fertilization programs.

Because the emphasis of *Cryopreservation and Freeze-Drying Protocols* is on methodology, it is our intention to provide readers with the tools to make practical progress without reference to other sources. Each chapter deals with an organelle, cell, or tissue type: a short introduction on the status of its biostorage development is followed by a detailed description of the materials required and a methodological protocol to be followed, with explanatory notes.

This is very much a first edition; we hope and trust that future editions will contain cryopreservation and freeze-drying protocols for cells, tissues, and organs that are at present still recalcitrant to successful preservation.

All of the protocols that follow are the result of painstaking development and modification over many years; we thank each contributor for providing a distillation of their efforts that should allow fresh application of their protocols in similar laboratories or units around the world.

John G. Day
Mark R. McLellan

Contents

Preface v
Contributors ix

CH. 1. Overview
    Mark R. McLellan and John G. Day 1

CH. 2. Virus Cryopreservation and Storage
    Ernest A. Gould 7

CH. 3. Freeze-Drying and Cryopreservation of Bacteria
    Stephen F. Perry 21

CH. 4. Freeze-Drying of Yeasts
    Sugio Kawamura, Yukie Murakami, Yukie Miyamoto,
    and Kazuo Kimura 31

CH. 5. Cryopreservation of Yeast Cultures
    Chris J. Bond 39

CH. 6. Cryopreservation and Freeze-Drying of Fungi
    Jacqueline A. Kolkowski and David Smith 49

CH. 7. Cryopreservation of Pathogenic and Nonpathogenic
    Free-Living Amoebae
    Simon Kilvington 63

CH. 8. Long-Term Cryopreservation of Thylakoid Membranes
    Dirk K. Hincha and Jürgen M. Schmitt 71

CH. 9. Cryopreservation of Algae
    John G. Day and Mitzi M. DeVille 81

CH. 10. Cryopreservation of Plant Protoplasts
    Brian W. W. Grout 91

CH. 11. A Two-Step or Equilibrium Freezing Procedure
    for the Cryopreservation of Plant Cell Suspensions
    Elly W. M. Schrijnemakers and Frank Van Iren 103

CH. 12. Vitrification of Plant Cell Suspensions
    Poula J. Reinhoud, Atsuko Uragami, Akira Sakai,
    and Frank Van Iren 113

CH. 13. Cryopreservation of Shoot-Tips and Meristems
    Erica E. Benson 121

Ch. 14.	Cryopreservation of Seeds	
	Hugh W. Pritchard	133
Ch. 15.	Cryopreservation of the Sperm of the Pacific Oyster *Crassostrea gigas*	
	Iain R. B. McFadzen	145
Ch. 16.	Cryopreservation of Fish Spermatozoa	
	Krishen J. Rana	151
Ch. 17.	Cryopreservation of Avian Spermatozoa	
	Graham J. Wishart	167
Ch. 18.	Cryopreservation of Animal and Human Cell Lines	
	Christopher B. Morris	179
Ch. 19.	Cryopreservation of Semen from Domestic Livestock	
	Mark R. Curry	189
Ch. 20.	Cryopreservation of Mammalian Embryos: *Slow Cooling*	
	Jocelyn E. Hunter	199
Ch. 21.	Cryopreservation of Mammalian Embryos: *Vitrification*	
	Magosaburo Kasai	211
Ch. 22.	Cryopreservation of Human Gametes	
	Jocelyn E. Hunter	221
Ch. 23.	Cryopreservation of Human Red Blood Cells	
	Michael J. G. Thomas and Susan H. Bell	235
Index		251

Contributors

SUSAN H. BELL • *Army Blood Supply Depot, Aldershot, Hants, UK*
ERICA E. BENSON • *Department of Molecular and Life Sciences, University of Albertay Dundee, Dundee, Scotland, UK*
CHRIS J. BOND • *National Collection of Yeast Cultures, AFRC Institute of Food Research, Norwich Laboratory, Colney, Norwich, UK*
MARK R. CURRY • *Royal Veterinary College, London, UK*
JOHN G. DAY • *Culture Collection of Algae and Protozoa, Institute of Freshwater Ecology, The Windermere Laboratory, Far Sawrey, Ambleside, Cumbria, UK*
MITZI M. DEVILLE • *Culture Collection of Algae and Protozoa, Institute of Freshwater Ecology, The Windermere Laboratory, Far Sawrey, Ambleside, Cumbria, UK*
ERNEST A. GOULD • *Institute of Virology and Microbiology, Natural Environment Research Council, Oxford, UK*
BRIAN W. W. GROUT • *Crisp Innovar Ltd., Glebe House, Reepham, Norfolk, UK*
DIRK K. HINCHA • *Institüt für Pflanzenphysiologie und Mikrobiologie, Frie Universität, Berlin, Germany*
JOCELYN E. HUNTER • *The Jackson Laboratory, Bar Harbor, ME*
MAGOSABURO KASAI • *Laboratory of Animal Sciences, College of Agriculture, Kochi University, Nankoku, Kochi, Japan*
SUGIO KAWAMURA • *Patent Microorganism Depository, National Institute of Bioscience and Human Technology, Higashi, Tsukuba, Ibaraki, Japan*
SIMON KILVINGTON • *Public Health Laboratory, Royal United Hospital, Combe Park, Bath, UK*
KAZUO KIMURA • *Patent Microorganism Depository, National Institute of Bioscience and Human Technology, Higashi, Tsukuba, Ibaraki, Japan*

JACQUELINE A. KOLKOWSKI • *Genetic Resources Collection, International Mycological Institute, Egham, Surrey, UK*
IAIN R. B. MCFADZEN • *Plymouth Marine Laboratory, Natural Environment Research Council, West Hoe, Plymouth, Devon, UK*
MARK R. MCLELLAN • *Acer Environmental, Acer House, Manor Park, Nr. Daresbury, Cheshire, UK*
YUKIE MIYAMOTO • *Patent Microorganism Depository, National Institute of Bioscience and Human Technology, Higashi, Tsukuba, Ibaraki, Japan*
CHRISTOPHER B. MORRIS • *European Collection of Animal Cell Cultures, PHLS Centre for Applied Microbiology and Research, Porton Down, Salisbury, Wiltshire, UK*
YUKIE MURAKAMI • *Patent Microorganism Depository, National Institute of Bioscience and Human Technology, Higashi, Tsukuba, Ibaraki, Japan*
STEPHEN F. PERRY • *Quality Assurance Laboratory, Central Public Health Laboratory, London, UK*
HUGH W. PRITCHARD • *Jodrell Laboratory, Royal Botanic Gardens, Kew, Ardingly, West Sussex, UK*
KRISHEN J. RANA • *Institute of Aquaculture, University of Stirling, Stirling, UK*
POULA J. REINHOUD • *Institute of Molecular Plant Sciences, Leiden University, Wassenaarsweg, Leiden, The Netherlands*
AKIRA SAKAI • *The Institute of Low Temperature Science, Hokkaido University, Sapporo, Japan*
JÜRGEN M. SCHMITT • *Institüt für Pflanzenphysiologie und Mikrobiologie, Frie Universität, Berlin, Germany*
ELLY W. M. SCHRIJNEMAKERS • *Institute of Molecular Plant Sciences, Leiden University, Wassenaarsweg, Leiden, The Netherlands*
DAVID SMITH • *Genetic Resources Collection, International Mycological Institute, Egham, Surrey, UK*
MICHAEL J. G. THOMAS • *Army Blood Supply Depot, Aldershot, Hants, UK*
ATSUKO URAGAMI • *Hokkaido National Experimental Station, Sapporo, Japan*
FRANK VAN IREN • *Institute of Molecular Plant Sciences, Leiden University, Wassenaarsweg, Leiden, The Netherlands*
GRAHAM J. WISHART • *Department of Molecular and Life Sciences, University of Albertay Dundee, Scotland, UK*

CHAPTER 1

Overview

Mark R. McLellan and John G. Day

1. Introduction

Nature dictates that biological material will decay and die. The structure and function of organisms will change and be lost with time, as surely in laboratory cultures as in the biologists who study and manipulate them. Attempts to stop the biological clock have been conjured by minds ancient and modern; at the heart of many such schemes have been experiments with temperature and water content.

Whereas refrigeration technology provides a means of slowing the rate of deterioration of perishable goods, the use of much lower temperatures has proved a means of storing living organisms in a state of suspended animation for extended periods. The removal of water from viable biological material in the frozen state (freeze-drying) provides another means of arresting the biological clock by withholding water, and commencing again by its addition.

Over 40 years have passed since the first demonstration of the effective cryopreservation of spermatozoa was made (1). The potential of storing live cells for extended, even indefinite, periods quickly caught the imagination of biologists and medics working in diverse fields, and experiments to cryopreserve many thousands of organelle, cell, tissue, organ, and body types have been, and continue to be, performed. Key milestones have been the successful cryopreservation of bull spermatozoa (2); the first successfully frozen and thawed erythrocytes (3); the first live birth of calves after insemination using frozen spermatozoa (4); successful cryopreservation of plant cell cultures (5); cryopreservation of a plant callus *(6);* the successful recovery of frozen mouse embryos *(7,8);* and the use of cryopreservation to store embryos for use in human *in vitro* fertilization programs *(9).* Furthermore, cryopreservation has become widely accepted as the optimal method for the preservation of microorganisms *(10–13).*

Cryopreservation and freeze-drying are widely employed to conserve microbial biodiversity *(11–13)* (*see* Chapters 2–7, and 9). This is one of the key roles performed by microbial service culture collections. More recently, cryopreservation has been accepted as an appropriate technique to preserve endangered plant *(14)* and animal *(15)* species (*see* Chapters 14 and 20). However, many cells and tissues for which there is a need for long-term biostorage await suitable methodologies. It is to be hoped that we are on the verge of cryopreserved transplant organs, frozen by vitrification; reproducible freezing of teleost eggs or embryonic stages; as well as the successful cryopreservation of human oocytes; a greater range of plant tissues; and a broader range of microalgae and protozoa.

A common misconception among noncryobiologists is that successful cryopreservation methods for one strain or species are transferable to similar cells or organisms. Although this is sometimes true, it is far from the rule. With different biology comes a different response to cryoprotectants and freezing; a preservation protocol may need adjustments, or to be constructed afresh for the material under study. It is worth a brief word on how such methods are determined.

It is usually the case that cryoprotectants must be added to protect cells during cooling, and careful manipulation of temperature excursion is required to control the size, configuration, and location of ice crystals. Therefore, choice and concentration of cryoprotectants, and rate of cooling must be optimized as the basis for any protocol. An accidental discovery was the spur for modern cryobiology *(1);* Polge's discovery of glycerol as an effective protectant allowed rapid advances in mammalian spermatozoa freezing. Dimethyl sulfoxide (DMSO), methanol, ethylene glycol, and hydroxyethyl starch (HES) have been added to the list of effective cryoprotectants. Many successful protocols have been developed empirically, by optimizing choice, concentration, time, and temperature of addition of cryoprotectant; along with the rate of cooling.

Much is known of the response of cells to low temperatures, and the effects of cryoprotectants, as a result of the efforts of scientists from a range of disciplines over the past 50 yr. The subject has its own considerable and complex literature to which the reader is referred for further information *(16–21)*. An outline of the major principles is given in the Introduction to Chapter 10. Such understanding has aided formulation of cryopreservation protocols by predicting optimum cooling rates from measured biophysical characters *(22)* and by direct visualization of cells and organisms during cooling *(23)*.

The formulation of freeze-drying protocols are as yet firmly empirically based; it has until recently been the case that the freeze-drying community have not accessed relevant information available from cryobiological studies. Further understanding of the effects of the sublimation phase of freeze-drying on cell biology is required, if techniques employed by microbiologists are to be extended to a range of eukaryotic cells, including erythrocytes and mammalian spermatozoa.

In order for a biostorage method to be acceptable as a routine laboratory practice, several criteria need to be fulfilled. Ideally, it should be relatively simple; complex procedures prior to freezing or freeze-drying may make the method more cumbersome or expensive than the culture methods it replaces. In addition, postthaw viability should be high, in order that cultures can regenerate rapidly, and preexisting freeze-resistant mutants are not selected. Many culture collections and gene banks insist on high recovery values prior to a protocol being adopted for regular use; 50% viability postthaw has been accepted in some culture collections as a nominal cutoff for adopting maintenance by cryopreservation alone *(24,25)*. Additionally, the storage method adopted should give level recovery rates with time; there is good evidence that a cryopreservation method yielding high initial recovery values, maintains viability at that level on prolonged storage *(26,27)*. The same may not be true of freeze-dried cultures or macromolecules, which are recommended to be stored at refrigerator or freezer temperatures.

As evidenced by the list of contributors to this volume, the cryobiological community embraces a wide range of specialists; medical scientists, plant-, animal, and microbiologists. Since much of the information on cryopreservation and freeze-drying is scattered, or bound in with theoretical literature, it is sometimes difficult to supply a recipe methodology for a particular purpose. We hope this handbook will be useful in providing clear and concise instructions for the long-term storage of a wide range of materials across the biological kingdoms.

References

1. Polge, C., Smith, A. U., and Parkes, A. S. (1949) Revival of spermatozoa after vitrification and dehydration at low temperatures. *Nature* **164,** 666.
2. Smith, A. U. and Polge, C. (1950) Storage of bull spermatozoa at low temperatures. *Vet. Rec.* **62,** 115–117.
3. Smith, A. U. (1950) Prevention of haemolysis during freezing and thawing of red blood cells. *Lancet* **ii,** 910,911.
4. Stewart, D. L. (1951) Storage of bull spermatozoa at low temperatures. *Vet. Rec.* **63,** 65,66.
5. Latta, R. (1971) Preservation of suspension cultures of plant cells by freezing. *Can. J. Bot.* **49,** 1253,1254.
6. Bannier, L. J. and Steponkus, P. L. (1972) Freeze preservation of callus cultures of *Chrysanthemum morifolium Ramat. HortScience* **7,** 194.
7. Whittingham, D. G., Leibo, S. P., and Mazur, P. (1972) Survival of mouse embryos frozen to −196 and −296°C. *Science* **178,** 411–414.
8. Wilmut, I. (1972) The effects of cooling rate, cryoprotectant agent and stage of development on survival of mouse embryos during freezing and thawing. *Life Sci.* **11,** 1071–1079.
9. Cohen, J., Simons, R., Fehilly, C. B., Fishel, S. B., Edwards, R. G., Hewitt, J., Rowland, G. F., Steptoe, P. C., and Webster J. M. (1985) Birth after replacement of hatching blastocyst cryopreserved at expanded blastocyst stage. *Lancet* **i,** 647.
10. Heckley, R. J. (1978) Preservation of microorganisms. *Adv. Appl. Microbiol.* **24,** 1–54.
11. Hatt, H. (ed.) (1980) *American Type Culture Collection Methods. I. Laboratory Manual on Preservation Freezing and Freeze-Drying*. ATCC, Rockville, MD.
12. Kirsop, B. E. and Snell, J. S. S. (eds.) (1984) *Maintenance of Microorganisms.* Academic, London.
13. Kirsop, B. E. and Doyle, A. (eds.) (1991) *Maintenance of Microorganisms and Cultured Cells.* Academic, London.
14. Withers, L. A. (1987) The low temperature preservation of plant cell, tissue and organ cultures and seed for genetic conservation and improved agriculture, in *The Effects of Low Temperatures on Biological Systems* (Grout, B. W. W. and Morris G. J., eds.), Edward Arnold, London, pp. 389–409.
15. Seymour, J. (1994) Freezing time at the zoo. *New Scientist* **No. 1910,** 21–23.
16. Grout, B. W. W. and Morris G. J. (eds.) (1987) *The Effects of Low Temperatures on Biological Systems.* Edward Arnold, London.
17. Morris, G. J. and Clarke, A. (eds.) (1981) *The Effects of Low Temperatures on Biological Systems.* Academic, London.
18. Ashwood-Smith, M. J. and Farrant, J. (eds.) (1980) *Low Temperature Preservation in Biology and Medicine,* Pitman Medical, Tunbridge Wells, Kent.
19. Franks, F. (1985) *Biophysics and Biochemistry at Low Temperatures.* Cambridge University Press, London.
20. Steponkus, P. L. (ed.) (1992) *Advances in Low Temperature Biology* vol. 1. JAI, London.

21. Steponkus, P. L. (ed.) (1993) *Advances in Low Temperature Biology* vol. 2. JAI, London.
22. Pitt, R. E. and Steponkus, P. L. (1989) Quantitative analysis of the probability of intracellular ice formation during freezing of isolated protoplasts. *Cryobiology* **26,** 44–63.
23. McGrath, J. J. (1987) Temperature controlled cryogenic light microscopy—an introduction to cryomicroscopy, in *The Effects of Low Temperatures on Biological Systems* (Grout, B. W. W. and Morris G. J., eds.), Edward Arnold, London, pp. 234–268.
24. Leeson, E. A., Cann, J. P., and Morris, G. J. (1984) Maintenance of algae and protozoa, in *Maintenance of Microorganisms* (Kirsop B. E. and Snell, J. S. S., eds.), Academic, London, pp. 131–160.
25. McLellan, M. R., Cowling, A. J., Turner, M., and Day, J. G. (1991) Maintenance of algae and protozoa, in *Maintenance of Microorganisms and Cultured Cells* (Kirsop B. E. and Doyle A., eds.), Academic, London, pp. 183–208.
26. McLellan, M. R. (1989) Cryopreservation of diatoms. *Diatom. Res.* **4,** 301–318.
27. Brown, S. and Day, J. G. (1993) An improved method for the long-term preservation of *Naegleria gruberi. Cryo-Lett.* **7,** 347–352.

CHAPTER 2

Virus Cryopreservation and Storage

Ernest A. Gould

1. Introduction

Viruses are noncellular forms of life and are much smaller and less biochemically complex than the simplest unicellular organisms. They consist of either RNA or DNA as a single molecule, or in some cases as a segmented genome, enclosed by one or more proteins. These proteins protect the nucleic acid from degradation; deliver it to the host cells that reproduce the virus; transcribe the nucleic acid (in the case of negative stranded genomes); and assist the virus to expose the nucleic acid to the biochemical machinery inside susceptible host cells. This relative simplicity has in part been the secret of the success of viruses in coexisting with all known life forms.

In general, DNA viruses are more stable than RNA viruses but both types are extremely stable and can be preserved relatively easily. Many viruses can be kept for months at refrigerator temperatures and stored for years at very low temperatures without the need for special preservatives or carefully regulated slow freezing techniques. Their simple structure, small size, and the absence of free water are largely responsible for this stability. Viruses with lipid envelopes are often less stable than non-enveloped viruses at ambient temperatures but survive well at ultra-low temperatures or in the freeze-dried state.

A variety of procedures exists for maintaining virus stocks and these depend to some extent on the peculiar properties of the particular viruses. Although the protocols in this chapter are devoted to cryopreservation and freeze-drying procedures, it is worth mentioning in general terms some recognized methods of maintaining viruses for relatively long periods without the need for specialized technical equipment (*see* Note 1).

Perhaps the most widely reported virus for long-term survival is the smallpox virus which is believed to be capable of surviving decades or possibly centuries in the dried form. Church crypts may contain infectious smallpox virus in the bodies of smallpox victims. Further examples of long-term survival include some tick-borne arboviruses. The ticks often have a very long life cycle, during which the virus remains viable. In some cases the virus is passed through the egg to the next generation. Under appropriate laboratory conditions the live infected ticks can be maintained for 1 or 2 yr. If the ticks are then allowed to take a bloodmeal, they develop through the next stage of their life cycle and remain infected. The virus can be retrieved at any time. These examples serve merely to illustrate the relative stability of some viruses and also the wide variety studied in the past 80–90 yr.

The infectious and often pathogenic nature of viruses means that they must be handled carefully by experienced personnel in purpose-designed and approved laboratories (*see* Note 2). In addition to the need for safe working practices as directed in the appropriate guidelines, work with viruses also requires the use of aseptic technique and an awareness of the risks of contamination either by other viruses or by other microorganisms. It is absolutely imperative that a virus that is being prepared for long-term storage and therefore as future reference material, should be handled in a virus-free environment. This can be achieved by various means but is most satisfactorily accomplished using a safety cabinet or laboratory that has been fumigated prior to the impending work. For tissue-culture work, it is also good practice to use sterile disposable pipets, and so on for all manipulations involving production of the virus stocks. Where plants, insects, or animals are involved in the virus production process, clean rooms (complying with all appropriate safety regulations) must be set aside before virus production and preservation commences. Clearly, there will be many instances where these conditions cannot be fulfilled precisely, for example, in diagnostic or research laboratories involving analysis of large numbers of field samples. In these situations it is good practice to preserve virus samples, in the first instance, with the minimum number of manipulations. Subsequent long-term preservation should then be performed on viruses isolated and amplified from the initially stored field samples.

There is an extensive literature on basic virology and the maintenance of viruses (1–4). The following protocols describe the preservation of a wide range of viruses and cover those routinely used at the NERC Institute of Virology and Environmental Microbiology, Oxford, UK.

2. Materials
2.1. Cryopreservation at 4°C (and –20°C)
1. 4°C Refrigerator and –20°C freezer.
2. Low- to medium-speed refrigerated centrifuge.
3. Sterile universal bottles (either glass or polypropylene).
4. Chloroform.
5. Aluminum foil.
6. Large plastic tray, metal gauze, cotton gauze, glass or Perspex lid to cover the tray, roll of plastic tape.
7. Anhydrous calcium chloride or silica gel.
8. Strong pair of scissors.
9. Plastic funnel that will fit into the neck of the universal bottles.
10. Sterile glass rod.
11. Facilities for culturing bacteriophage.
12. Virus-infected leaves.

2.2. Cryopreservation at –70°C
1. Ultra-low temperature freezer capable of maintaining a temperature of –70°C or lower.
2. Screw-capped cryotubes—preferably small volumes, such as 1- or 2-mL capacity (Nunc, Roskilde, Denmark).
3. Ice bath.
4. Sterile Pasteur pipets or sterile disposable graduated pipets.
5. Cold-protective gloves.
6. Indelible marker pen.

2.3. Cryopreservation at –70°C
1. Phosphate-buffered saline (PBS) at pH 7.4–7.6 containing 10% fetal bovine serum and 200 U/mL of penicillin G and 200 µg/mL of streptomycin sulphate.
2. Universal plastic bottles.
3. Sterile graduated pipets.
4. Sterile mortar and pestle, or sterile ground glass homogenizers or Waring Blender.
5. Ice bath.
6. Sterile cryotubes.

7. Low-speed refrigerated centrifuge.
8. Virus in intact arthropods, animal tissue specimens or plant tissue.
9. Acetone.
10. Cold-protective gloves.

2.4. Cryopreservation at −70°C

1. Sterile glass universal bottles (flat-bottomed) containing 6-mm sterile glass beads (approx 8–10/bottle).
2. Low-speed refrigerated centrifuge.
3. Small volume cryotubes (*see* Section 2.2., item 2).
4. Sterile graduated pipets or Pasteur pipets.
5. Brains from virus-infected mice.
6. Small vortex mixer, e.g., Whirlimixer or equivalent.
7. Sterile PBS.
8. Cold-protective gloves.

2.5. Cryopreservation in Liquid Nitrogen

1. Heat shrink cryotubing, e.g., Nunc Cryoflex or equivalent.
2. Screw-capped cryotubes: preferably small volume (1 or 2 mL).
3. Liquid nitrogen storage tanks.
4. Protective gloves and face mask.
5. Indelible marker pen.
6. Bunsen burner.
7. Thermos flask containing liquid nitrogen.

2.6. Freeze-Drying Viruses for Long-Term Preservation

1. Glass freeze-drying ampules: These come in various sizes. For most purposes, ampules of approx 2 or 5 mL capacity are satisfactory.
2. Air/gas torch, producing a narrow flame, preferably with a two-sided outlet to provide heat on two sides of the glass ampule at the same time. This is not absolutely essential but does simplify the procedure of sealing ampules.
3. Long forceps.
4. Aluminum foil.
5. Sterilizing facilities: Autoclave or drying oven.
6. Freeze dryer with condensing chamber, high performance diffusion (vacuum) pump, and branched manifold attachment suitable for connecting the ampules individually.
7. Sterile Pasteur pipets.
8. Protective gloves and face shield.
9. Thermos flask and either a mixture of dry ice/methanol or liquid nitrogen.
10. Good quality sticky cloth tape and indelible marker pen.

Virus Cryopreservation and Storage

11. High voltage spark tester (not essential).
12. High vacuum grease.

3. Methods

There are many different viruses but in general the principles and practices that are described should apply to most of them. The most important ground rules to remember are:

1. Viruses are hazardous, therefore handle them in purpose-designed facilities with appropriate safety procedures (*see* Note 2).
2. Keep virus preparations at 4°C when they are not being used or preserved long term.
3. Unless it is necessary to reduce infectivity for a scientific purpose, maintain only high titers of virus.
4. Freeze and thaw viruses rapidly and infrequently!
5. Unless it is required for a specific purpose, do not subculture viruses unnecessarily.
6. The lower the temperature the longer the virus will survive (*see* Notes 3 and 4).
7. If possible, virus stocks for long-term preservation should be backed up by storage in more than one location.

3.1. Cryopreservation at 4°C (and –20°C)

3.1.1. Bacteriophage

Most bacteriophage can be stored at 4°C for a few years. The infectivity will decrease slowly with time but it is usually a simple task to revitalize the stock after 1 or 2 yr by culturing the phage in the appropriate bacterial host.

1. Culture the bacteriophage, preferably under one-step growth conditions to yield a high infectivity titer (probably $>1 \times 10^9$ PFU/mL).
2. Clarify the infective culture medium by centrifugation for 20 min at 3000g.
3. Store the supernatant medium in either a screw-capped glass or in polypropylene sterile universal bottles at 4°C, which should be wrapped in silver foil to protect the contents from the light.
4. Add 2 or 3 drops of chloroform (assuming the bacteriophage do not have lipid envelopes) to each bottle to ensure sterility (*see* Notes 5 and 6).

3.1.2. Baculoviruses

The more complex viruses, such as baculoviruses or pox viruses can also be stored at 4°C for a few years, and they also preserve satisfactorily at –20°C.

1. Culture the virus in appropriate cells using a relatively low multiplicity of infection, i.e., 0.1–0.01 infectious virions per cell.

2. Collect the infectious supernatant culture medium after incubation for 48–72 h (for baculoviruses and animal pox viruses) at the appropriate temperature. The objective is to obtain high titer preparations, therefore it is a good idea to optimize the culture conditions before preparation of the virus stocks for preservation. Usually, these viruses produce marked cytopathic effects on the infected cells, causing infectious virus to be released into the supernatant culture medium as the cells are killed and lysed.
3. Clarify the supernatant medium by centrifugation at 2000g for 10 min.
4. Store the clarified medium in sterile plastic screw-capped bottles at 4°C out of the light. The virus will preserve equally well at –20°C.
5. If baculovirus-infected caterpillars are available, either from field sources or from caterpillars reared in the laboratory, they can be placed directly into bottles and stored at –20°C for years. The virus infectivity will decrease only slightly.
6. Thaw the frozen virus samples rapidly by placing the cryotubes in a water bath at 37°C. Thawing should be carried out just before the virus is to be used unless it is known that the virus has good thermostability characteristics when held at laboratory temperatures. Remove the cryotubes from the water bath immediately after thawing is completed and incubate at 4°C until they are required (*see* Notes 5 and 6).

3.1.3. Plant Viruses

Many plant viruses can also be stored at 4°C for a few years if the infected plant tissue is dehydrated chemically.

1. Place sufficient anhydrous calcium chloride to cover the bottom of the tray and cover it with a metal gauze or screen. The ends of the gauze should be bent over to create a platform over the calcium chloride. Place cotton gauze over the platform.
2. Collect infected leaf tissue and cut it into small pieces approx 0.5 in. square, avoiding the thick stems and ribs of the leaves. Distribute the cut pieces of leaves onto the cotton gauze. It is a good idea to work aseptically and to ensure that material containing other plant viruses is not nearby.
3. Cover the tray with a suitable piece of glass or plastic and seal it to the tray with plastic tape (electrical insulation tape is ideal).
4. Place in a refrigerator for approx 8 d.
5. The virus can be stored in small dry glass or plastic bottles (about 25–30 mL capacity, with a wide neck of approx 2.5-cm diameter) containing a dehydrant, such as silica gel. For convenience, the silica gel can be prepared as small packets in cotton gauze tied with thread (if not available as commercially supplied packs). On d 7, prepare the silica gel packs and

place in a drying oven at about 60°C overnight. The silica gel will turn pink when dry (blue when not dry). Alternatively a piece of dried calcium chloride (approx 100 mg) can be placed into the bottle.
6. On d 8, allow the silica gel packs to cool to room temperature and then aseptically place one at the bottom of each storage bottle.
7. Remove the tray containing the dried leaves from the refrigerator and place at room temperature for 1 h to equilibrate then transfer the dried infected pieces of leaf to the bottles containing the silica gel. The simplest method of transferring the leaves to the bottles is to pour them down a large-necked funnel directly into the bottles. A sterile glass rod or pipet can be used to push the pieces of leaf tissue into the bottles.
8. Seal the bottle immediately with a screw cap and wrap 3 or 4 layers of plastic tape tightly around the joint of the cap with the bottle as additional security against water vapor entering the bottle.
9. Place the bottles in the refrigerator, preferably protected from exposure to the light (*see* Notes 5 and 6).

3.2. Cryopreservation at –70°C

Each of the aforementioned methods described for preservation at –20°C is equally applicable for preservation at –70°C (*see* Note 7).

1. Before virus is harvested from the cells, label the cryotubes in which it will be preserved, paying attention to the advice on record keeping (*see* Note 3). For a working stock of virus we routinely label 50 × 2 mL sterile screw-capped cryotubes (available from all well-known suppliers of tissue-culture plasticware).
2. Place the flask containing the clarified virus suspension, either as tissue-culture supernatant medium or as cell lysate in tissue-culture medium (*see* Note 8), in an ice bath for a few minutes (*see* Notes 9–11).
3. Dispense small volumes (from 0.1–1 mL) of the clarified medium aseptically into the cryotubes using a sterile pipet and ensuring that the cap of each cryotube is screwed down firmly. (Usually dispense 0.2-mL aliquots from a 10-mL disposable pipet [*see* Note 12].)
4. Wearing protective gloves to handle the trays and racks in the freezer, transfer the cryotubes containing the dispensed virus directly to the –70°C freezer (*see* Notes 13 and 14). We use storage racks with trays that contain partitions suitable for cryotubes up to 2-mL capacity.
5. Virus required for experimentation should be obtained from the rack, recording precisely which cryotube was removed. It should be placed in a water bath at 37°C immediately and removed as soon as it has thawed. Use the virus as soon as possible after thawing keeping it at 4°C until used.

3.3. Cryopreservation at −70°C

Viruses present in infected arthropods, animal tissues or plants that need to be preserved at ultra-low temperatures can all be frozen directly or they can be prepared as clarified suspensions in diluent and then frozen in a similar way to that described in Section 3.2., step 4. The precise technical procedures differ slightly as indicated:

1. Place infected arthropods (usually held in a plastic bottle) at −70°C until frozen to kill them and to soften the tissue.
2. Prepare pools of the arthropods (usually the same species in each pool, consisting of up to 100/pool) and suspend the pools in the PBS at the rate of up to 20 arthropods in 1 mL of buffer (*see* Note 8). Put the suspension into a sterile glass homogenizer or a sterile mortar that is cooled on an ice bath, and grind the specimens until the arthropods are totally disrupted (*see* Note 15).
3. Clarify the suspension by centrifugation at about 2000*g* for 20 min at 4°C and then dispense small volumes into cryotubes, record, and freeze as described in Section 3.2., step 4.
4. For recovery of viruses use method detailed in Section 3.2., step 5.

3.4. Cryopreservation at −70°C

Another method favored by virologists working with arboviruses relies on glass beads to release virus from mouse brain cells or other relatively soft tissue. Very young mice inoculated intracerebrally or intraperitoneally with arboviruses produce high virus infectivities in the brain. When the mice are sick they are killed using terminal anesthesia, and the brains are removed aseptically for processing *(2)*.

1. Aseptically remove the infected mouse brains from the mice at the appropriate time after infection with the arbovirus and place them in sterile universal glass bottles containing 6-mm sterile glass beads.
2. Place no more than 10 newborn mouse brains into one glass flat-bottomed universal bottle containing about 8–10 glass beads and screw the cap on tightly.
3. Vortex the brains for 1 min (in a safety cabinet).
4. Add 2 mL of cold (4°C) PBS per brain, replace the cap tightly, and repeat the vortexing procedure (*see* Note 8).
5. Centrifuge the suspension at 2000*g* for 10 min.
6. Dispense the clarified supernatant medium in properly labeled cryotubes, replace the cap securely, and freeze at −70°C as described in Section 3.2., step 4.
7. For recovery of viruses use method detailed in Section 3.2., step 5.

Virus Cryopreservation and Storage

3.5. Cryopreservation in Liquid Nitrogen

1. Cut a length of Cryoflex tubing sufficient to extend 2 cm beyond each end of the cryotube.
2. Dispense viruses into cryotubes as detailed in Section 3.2., step 3.
3. Insert the correctly labeled cryotube containing the virus in the center of the cut length of Cryoflex tubing.
4. Heat the tubing gently using the flame from a Bunsen burner or heat gun. The heat will shrink the tubing around the cryotube. Note: It is not necessary to heat the tubing to a high temperature (*see* Note 16).
5. Reheat the ends of the tubing and squeeze or crimp the ends with a large pair of forceps (or equivalent) to provide a seal. The ends of the Cryoflex tubing can be melted to ensure an absolute seal.
6. Snap freeze the sealed cryotubes in a small volume of liquid nitrogen in a thermos flask (wear a face mask and gloves).
7. Place the frozen sealed cryotubes into the appropriate compartments of the liquid nitrogen tank and keep a detailed record of the position, experiment number, date, and so on, of the samples (*see* Note 17).
8. When the frozen virus is required for experimentation, remove the cryotube from the nitrogen, thaw it at 37°C for the minimum time necessary, and use a scalpel blade to cut the Cryoflex tubing at the position of the silicone gasket on the cap of the cryotube. Unscrew the cap with the Cryoflex tubing still attached to it.

3.6. Freeze-Drying Viruses for Long-Term Preservation

This is probably the most satisfactory method of preserving viruses for very long periods. There are several variations in the technical procedures depending on the specific design of the freeze-drying equipment. For small numbers of samples and small volumes of virus, the simplest and most effective method involves only one vacuum stage because the glass ampules are placed directly onto the branched exhaust manifold of the freeze-dryer.

1. Heat the neck of each ampule (about 2 cm from the top) by rotating it in the flame of a purpose-designed gas torch that presents the flame on both sides of the glass simultaneously. As the glass softens in the flame it will naturally push the glass inward. At this moment, using a pair of blunt forceps, gently stretch the neck of the ampule just sufficiently to cause a slight narrowing at the softest point. Do not stretch the neck more than about 5–10 mm. Remove the ampule from the flame as you stretch the neck and quickly roll it on a flat heat resistant surface to ensure it is reasonably

straight. Prepare a large number of ampules in this way as they can be stored indefinitely.
2. Place a piece of aluminum foil (two layers thick) over the end of each ampule and affix it to the shoulder of the ampule with a small piece of autoclave (heat resistant) paper tape.
3. Sterilize the ampules. We use dry heat, but autoclaving is also suitable.
4. Using a long thin Pasteur pipet or other equivalent applicator, carefully insert a small volume of the virus suspension into each ampule, ensuring that the volume of the sample is less than one-third of the ampule volume. We use 0.5 mL of sample in 2-mL ampules. When inserting the sample try to avoid contaminating the neck of the ampule with the virus.
5. Wearing protective gloves and a face shield, shell-freeze the virus suspension by vigorously rotating the ampule in a dry ice alcohol bath or in liquid nitrogen held in a wide-neck thermos flask. Once the sample is frozen, keep it frozen in a suitable container until all the other ampules are similarly frozen. Note that it is important to snap-freeze the sample around the surface of the ampule, hence the term shell-freeze. This helps to maintain a high infectivity by increasing the speed of freezing and drying.
6. Switch on the freeze-dryer about 30 min before it is to be used to ensure that the temperature of the condenser has reached at least $-40°C$.
7. Place a small amount of high vacuum grease on the manifold gaskets and switch on the diffusion pump. Use empty ampules to seal off the ports not required. Arrange them on the manifold so that the number of available ports exactly matches the number of samples to be attached.
8. Immediately load the frozen ampules onto the branched exhaust manifold. Perform this operation as quickly as possible (*see* Note 18).
9. The vacuum will start to develop as soon as the last ampule is connected to a spare port on the branched exhaust manifold.
10. Normally, the samples on the manifold will remain frozen because the vacuum is generated quite quickly. The freeze-drying process should be allowed to take place until the samples are completely dry, at which time there will be no moisture of condensation on the outside of the ampules. With small samples and low numbers of ampules, i.e., 5–10, the process should not take more than 3–4 h, although convenience samples should be dried overnight.
11. When the samples are dry, seal the ampules under vacuum at the narrow point of the neck, which was prepared earlier. Use a suitable gas torch to melt the glass at the constriction. Allow the glass to separate as each end seals itself. Do not pull the glass ampule away as the glass melts. Once separated, use the flame of the torch to melt the top of the ampule so that it forms a thick and, therefore, strong seal.

12. If available, a high voltage spark tester can be used to test the integrity of the vacuum, but this is not absolutely essential (*see* Notes 19 and 20).
13. Label the ampules in such a way that they can be identified many years later. White cloth tape is ideal for this purpose.
14. Store the ampules at 4°C or lower if possible and avoid direct exposure to light (*see* Note 21).
15. After storage for a few days, open one of the ampules in an appropriate safety cabinet to test the infectivity of the virus. The ampules are designed to break at the neck. Place a triple-layered piece of alcohol-soaked paper toweling around the ampule and, wearing protective gloves, snap off the neck of the ampule while it is held inside the paper soaked in alcohol (alternative virucides are equally suitable). Reconstitute the contents of the ampule using sterile distilled water to the volume of the original starting material.
16. Check the infectivity of the virus in the test ampule by titration; freeze-drying should not significantly reduce infectivity. Test another ampule after 6 mo storage. If the titer of the virus has not altered significantly, the virus in the remaining ampules should remain viable for many years.

4. Notes

1. Many plant viruses are extremely stable in the dried form at room temperature, although low temperatures are preferable for longer term storage. Dried virus-infected leaves, placed out of the light, can be maintained for months or even years. Plant viruses in seeds will also survive long periods of storage. Some plant viruses establish long-term infections in plants or trees and this principle can be exploited to preserve the virus. Bacteriophage, i.e., viruses that infect bacteria, are usually stable for several years if kept at 4°C in the clarified nutrient broth used to grow the bacteria. Baculoviruses, i.e., viruses that infect insects, have been known to survive up to 40 yr in soil.
2. Most developed countries outside the former USSR have produced Approved Codes of Practice for work with pathogenic microorganisms and in Europe they have now been incorporated into the Regulations for Control of Substances Hazardous to Health (COSHH). Thus, one is legally required to conform to the standards recommended in the Codes of Practice. Before commencing any work with infectious viruses, the Hazard Grouping of the virus must be checked and all work must then be carried out under the appropriate conditions. In the United Kingdom, the Health and Safety Executive, Library and Information Services (Baynard's House, 1 Chepstow Place, Westbourne Grove, London. W2 4TF) advises on microorganisms hazardous to humans and the Ministry of Agriculture, Fisheries and Food (Hook Rise South, Tolworth, Surbiton, Surrey KT6 7NF, UK) advises on viruses hazardous to animals and plants.

3. The importance of good quality record keeping is often overlooked even by experienced scientists. Since the samples are likely to be kept for many years, it is absolutely essential that a precise record of all details is kept in a good quality book or card system. A computer record is also useful but there needs to be some degree of certainty that the data will be accessible many years later when the computer will have been replaced. It is strongly recommended to prepare a detailed label on good quality tape, written (or typed) in indelible ink.
4. Viruses can be preserved for long periods as nucleic acid. The purified nucleic acid of positive-stranded RNA viruses (i.e., those in which the viral RNA is the messenger RNA) and many DNA viruses (i.e., those that do not enclose essential enzymes in their structure) is infectious. This principle can be utilized to preserve these viruses for very long periods of time. The ethanol-precipitated RNA and DNA can be stored almost indefinitely at 4°C (or lower temperatures) under ethanol. The ethanol is important for long-term storage of RNA to inhibit enzymes that breakdown RNA. DNA can be stored either under ethanol or as dried DNA. This method of virus preservation is one of the most effective available but is not very widely used. Virus frozen as nucleic acid can probably be preserved almost indefinitely, and since it can be stored in extremely small volumes, many samples can be maintained without the need for large volumes of storage capacity.
5. If retention of virus infectivity is not essential, for example in cases where the sample is required as an antigen in an enzyme-linked immunosorbent assay (ELISA), it can be stored for many years at –20°C without loss of antigenic activity, even though the infectivity might be significantly reduced.
6. Long-term storage at –20°C of acetone-fixed virus infected cells on glass coverslips is a very convenient method of retaining specific virus antigens for serodiagnostic purposes.
7. Dry ice should only be used to preserve viruses in totally sealed containers. The optimal pH for virus storage is between pH 7.0 and 8.0. Viruses are relatively labile at pH 6.0 or below. It is therefore unwise to store virus preparations in unsealed containers on dry ice since the released carbon dioxide is absorbed through the joint between the cap and the cryotube, and the absorbed carbon dioxide reduces the pH of the preserved virus suspension.
8. Proteins in the form of serum or other biological material, in buffered isotonic salt solution or tissue-culture medium, can be used to preserve infectivity of most viruses held at ultra-low temperatures. The precise mechanisms of the protective effects are not known, but the proteins possibly provide buffering capacity against pH changes, assist in colloidal dispersion of the virus particles, and reduce or inhibit other processes that damage nucleic acids. Viruses contained in serum or tissues from human

or animal specimens can be stored at ultra-low temperatures without further treatment.
9. Virus preparations to be preserved from tissue-culture monolayers, cell suspensions, or allantoic fluid from infected fertile hen eggs should be clarified by centrifugation at about 2000g for 20 min at 4°C. The clarified preparation should then be dispensed and frozen immediately.
10. It is good practice to determine when maximum infectivity is produced in the culture and to harvest the virus at this time. Unless they are known to be very stable, viruses should not be held at room temperature for more than a few minutes.
11. Many viruses produce marked cytopathic effects and are efficiently released into the supernatant medium, others are less cytopathic and therefore retained within the infected cells. With released viruses, it is a simple matter to clarify the supernatant medium by sedimenting the cell debris. With viruses that remain in the cells, these should be harvested at the optimal time of virus production and lysed either by rapid freeze–thawing cycles, using a mixture of methanol and dry ice, or by ultrasonication at 4°C for 15 s (carrying out these manipulations according to the advice given in the recommendations of the Advisory Committee for Dangerous Pathogens, from the Health and Safety Executive, *see* Note 2). The lysed cell debris can then be removed and virus collected as clarified medium.
12. Viruses should be frozen rapidly and this is most readily accomplished by storing only small volumes (0.1– 0.5 mL) of virus suspension. Rapid freezing and thawing or reconstitution of a virus preparation is less harmful to the virus than slow freezing, thawing, and reconstitution. Moreover, for most research and diagnostic purposes, it is important to be able to reproduce the same result many times. By dispensing small aliquots, large numbers of samples of the same preparation can be stored in a freezer, each available to reproduce the same performance.
13. Virus infectivity is retained well at temperatures below –60°C. Many freezers, which are now available, can reliably maintain these ultra-low temperatures. In many virology laboratories, –70°C (or, more recently, –80°C) is the favored temperature, partly because viruses are known to survive for decades at –70°C and partly because modern freezers do not have to work at their maximum capacity to maintain this temperature, thereby increasing their reliability.
14. It is very important to ensure that the freezer has an alarm to warn if the freezer fails, i.e., if there is a rise in temperature of more than 5°C. Virus infectivity is significantly reduced if there is a slow rise in temperature. Ideally, a backup freezer should be available; some companies will supply one in emergencies.

15. The mortar and pestle method of extracting viruses from plant or animal tissue is widely used, although a Waring Blender is sometimes used with plant tissue, which is suspended in acetone. The acetone is then removed by evaporation and the dried precipitate is dispensed and frozen.
16. It is recommended to use only small volume cryotubes for storage of virus in liquid nitrogen. Each cryotube must be sealed in special tubing (Cryoflex Nunc or equivalent) to avoid the risk of cross contamination of viruses and also exposure of the operator to virus-containing aerosols when cryotubes are removed from the nitrogen.
17. It is also important to remember that liquid nitrogen storage tanks have to be checked and replenished with liquid nitrogen regularly. Modern equipment is often fitted with a self-filling device from a reservoir.
18. In order to minimize potential cross-contamination, never freeze-dry more than one virus at a time.
19. Clean the ampule attachment points on the manifold with 70% (v/v) ethanol prior to the next usage and smear a very small amount of vacuum grease onto each ampule attachment point.
20. On completion of the freeze-drying process, when the freeze-dryer has reached room temperature, wipe the condenser chamber several times with a suitable virucidal agent to ensure there is no viable virus present.
21. Freeze-dried preparations of virus can be maintained for decades at 4°C. Storage of samples in the dark and at lower temperatures increases the shelf-life. Although this principle has not been tested exhaustively for every known virus, it has been demonstrated with very many different viruses.

References

1. Kurstak, E. (ed.) (1991) *Viruses of Invertebrates.* Dekker, New York.
2. Mahy, B. W. J. (ed.) (1985) *Virology: A Practical Approach.* IRL, Oxford, Washington, DC.
3. McKinney, H. H. and Silber, G. (1968) Methods of preservation and storage of plant viruses, in *Methods in Virology,* vol IV (Maramarosch, K. and Koprowski, H., eds.), Academic, London, pp. 491–501.
4. Ward, T. G. (1968) Methods of storage and preservation of animal viruses, in *Methods in Virology,* vol IV (Maramarosch, K. and Koprowski, H., eds.), Academic, London, pp. 481–489.

CHAPTER 3

Freeze-Drying and Cryopreservation of Bacteria

Stephen F. Perry

1. Introduction

Freezing and freeze-drying techniques have become standard methods for the long–term maintenance of bacterial cultures. Both methods of preservation provide varying degrees of success with different species of bacteria, and neither technique results in 100% recovery of preserved cells *(1)*. There is no universally applicable method for the successful preservation of all bacteria, and where it is vitally important that cultures are not lost, it is advisable to use both methods in parallel.

1.1. Freeze-Drying

Simple freezing and freeze-drying regimes are often established empirically. However, it is possible to apply scientific principles to the control of parameters allowing the optimization of processes for the freezing and drying of organisms *(2)*. Thus, heat and vapor transfer can be manipulated to maintain sublimation under optimal conditions of temperature and time.

Freeze-drying is a process in which frozen material is dried through the sublimation of ice *(3)*. The procedure consists of the following three stages: freezing, sublimation and desorption. Initially the material is frozen, causing a physical separation of the water as ice from the solids. In the second stage of the process, the ice is removed from the product by direct conversion to vapor (sublimation). To accomplish this transformation, energy is required in the form of heat. For sublimation to occur at the ice interface, the energy required for the solid–vapor transformation must be transported through the sample to the interface, requiring a temperature difference (usually termed a temperature gradient) between the heating source and the interface. The energy input must be controlled so that the quantity of vapor produced can be removed quickly enough to avoid conditions that contribute to structural breakdown (collapse), especially at the sublimation interface. After the vapor has been formed at the interface, it must be transported away from the sample. The removal of vapor requires mass transport, and necessitates a pressure difference, usually termed a pressure gradient, between the interface and the refrigerated condenser surface. A chemical desiccant, such as phosphorus pentoxide can be used to trap the small amounts of water involved, but it is more convenient to use a refrigerated condenser at –50°C.

Following the removal of ice crystals, what remains of the product is a concentrated solute phase which will become, at the end of the process, the freeze-dried material. The solute phase will still contain a significant quantity of strongly bound unfrozen water *(4)* (generally about 25–30 g water per 100 g solids) *(5)*. Most bacteria will not be structurally or chemically stable unless most of this bound water is removed during freeze-drying. The removal of this bound water is achieved through desorption. As in the other two stages of freeze-drying, it is necessary during desorption to input energy to form water vapor from the bound water molecules.

Two types of commercial freeze-dryer, the centrifugal and shelf, are in common use. In the former, freezing is brought about by evaporation that occurs when the vacuum is applied, and the cell suspension is centrifuged during initial freezing to increase the surface area and prevent frothing. For large culture collections, the centrifugal method has advantages in minimizing the likelihood of cross-contamination as ampules may be plugged after filling and sealed under vacuum on a manifold at the end of the secondary drying stage. However, for the inexperienced and infrequent user, centrifugal freeze-drying is more technically demanding. The method described in this chapter is specifically for shelf freeze-drying; methods for centrifugal drying are described elsewhere *(6)*. The initial decrease in numbers of viable cells during the drying process by either method generally is low. Shelf-life following centrifugal drying with heat sealing of ampules has been documented as greater than 35 yr *(7)* for some species. In my experience with medically important bacteria, survival following shelf-drying is several years; information on long-term stability is still lacking.

1.2. Cryopreservation

With cryopreservation, water is made unavailable to the bacteria by freezing, and the dehydrated cells are stored at low temperatures. Methods can be broadly classed according to the storage temperature; −20 to −30°C is achievable with standard laboratory freezers, −70°C with ultra-low temperature freezers, and −140 to −196°C with liquid nitrogen. Storage of cells in the nitrogen vapor phase (−140°C) or the liquid nitrogen phase (−196°C) is increasingly being used. At such low temperatures, cellular viability is almost independent of the period of storage, and biological systems are believed to be genetically stable *(8)*. Storage of cultures in the range of −60 to −80°C will often result in good viability and may be used when liquid nitrogen is not available or in noncritical applications where some loss of culture viability can be tolerated. Freezers operating within this temperature range are readily available and this method eliminates the need for a constantly available nitrogen supply. In general, temperatures above −30°C give poor results because of the formation of eutectic mixtures and hence the exposure of cells to high salt concentrations. Freezing removes the available free water, and in biological systems only a proportion of the total water is converted to ice. The removal of water by freezing increases the concentration of solutes in the remaining aqueous phase thus lowering the freezing point. As the temperature is further reduced, more ice forms and the residual solution becomes increasingly concentrated *(9)*. The damaging effects of freezing and thawing are believed to be associated with this formation of concentrated solutions as there is no evidence of mechanical injury to cells by extracellular ice *(10)*. The incorporation of a nonionic component, e.g., glycerol as a cryoprotectant, reduces the amount of ice at any temperature during cooling, thereby reducing the increase in ionic concentration.

To reduce the damage caused to cells by repeated freezing and thawing when subcultures are required, a method based on freezing bacterial suspensions with a cryoprotectant in the presence of glass beads has been devised *(11)*. This technique allows individual beads to be removed from the cryotube without thawing the whole sample. The method has proven to be a reliable and simple process requiring no further manipulation during storage.

2. Materials
2.1. Freeze-Drying

1. Suspending fluid: Inositol serum nutrient broth *(12)*, 6.67 g meso-inositol (Sigma, Poole, Dorset, UK), 100 mL sterile horse serum (Advanced Protein Products Ltd., Brockmoor, West-Midlands, UK), 0.825 g nutrient broth powder No. 2 (Unipath, Basingstoke, Hampshire, UK), 33.3 mL distilled water. The nutrient broth inositol and water are mixed thoroughly in a 250-mL conical flask, when the inositol has dissolved, the mixture is sterilized by autoclaving at 121°C for 15 min. When the broth mixture has cooled, the sterile horse serum can be added. The resulting suspension is distributed aseptically in 5-mL aliquots into bijoux bottles and incubated at 30°C for 2–3 d as a sterility check. The broth can be stored at –30°C and thawed when required.
2. Vials: Before use, 16 × 36 mm borosilicate neutral glass vials (Schubert Seals, VN1595, Gosport, Hampshire, UK) are heat sterilized for 2 h at 160°C. The vials may be regarded as clean when received as they are hermetically sealed by the manufacturer soon after cooling from 650°C. The vials can be labeled by inserting strips of Whatman (Maidstone, Kent, UK) No. 1 filter paper, with the culture identification number typed on, prior to sterilization.
3. Bungs: Chlorobutyl cruciform bungs (type 20133A, Schubert Seals) are heat treated to drive off moisture that cannot otherwise be removed during freeze-drying. Bungs are placed one-layer deep in a stainless steel tray, and heated in a hot air oven for 2 h at 110°C. The lid of the tray is removed when the oven is switched on and replaced at the end of the cycle (*see* Note 1).
4. Freeze-dryer (Edwards Freeze-dryer, Modulyo, Crawley, Sussex, UK).
5. Sterile plastic Pastets (Alpha Laboratories, Ltd., Eastleigh, Hampshire, UK).
6. Aluminum tear off caps (Shubert Seals).

2.2. Cryopreservation

1. Media: The suspending fluid for aerobic bacteria is 2.5 g nutrient broth powder No. 2 (Unipath), 15 mL glycerol (Sigma), 85 mL distilled water. The glycerol nutrient broth is dispensed into 5-mL aliquots and sterilized by autoclaving at 121°C for 15 min.

    For anaerobic bacteria, BGP medium *(13)* without the agar but with an additional 15% (v/v) glycerol is used: 1.0 g Tryptone (Unipath), 0.5 g NaCl, 0.3 g beef extract, 0.5 g yeast extract, 0.04 g cysteine hydrochloride, 0.1 g glucose, 0.4 g Na_2HPO_4, 15 mL glycerol, 85 mL distilled water. The medium is dispensed in 10-mL aliquots into universal bottles and steril-

ized by autoclaving at 121°C for 15 min. Both media can be stored at 4°C for several months prior to use.
2. Glass beads: Glass 2-mm diameter embroidery beads (Creative Beadcraft Ltd., Amersham, Buckinghamshire, UK) prewashed in tap water (*see* Section 3.1., step 1).
3. Sterilized vials: 20–30 of the prepared beads are placed in 2-mL screw-capped cryotubes (Nunc, Paisley, Scotland). The vials are capped and sterilized by autoclaving at 121°C for 15 min. Sterilized vials can be stored at room temperature until required.
4. Freezer precooled to –70°C.
5. Sterile plastic Pastets.

3. Methods
3.1. Freeze-Drying

Figure 1 depicts a flowchart showing the stages involved in the preparation of cultures prior to freeze-drying.

1. Prewash glass beads in tap water with a detergent (e.g., Flow Laboratories, Thame, Oxfordshire, UK, 7X phosphate-free laboratory detergent), followed by an acid ($0.1M$ HCl) wash to neutralize alkaline. Wash the beads repeatedly in tap water until the pH of the wash water is that of tap water. Then rinse with distilled or deionized water, and dry in a hot air oven.
2. Grow the bacteria on an appropriate nonselective, solid medium, such as nutrient or blood agar under the optimum growth conditions (*see* Notes 2 and 3).
3. Aseptically add 2 mL of the suspending fluid to the agar slope.
4. Suspend the culture using a sterile plastic loop and thoroughly mix the resulting cell suspension (10^8–10^{10} cells/mL).
5. Transfer the cell suspension to the remaining 3 mL of inositol broth and again mix thoroughly (*see* Note 4).
6. Using a sterile plastic Pastet, transfer approx 0.5 mL of the cell suspension to the bottom of a prelabeled vial (*see* Note 5). Care must be taken to ensure that the sides and top of the vial are not contaminated with the suspension. Insert the vial bungs halfway into the vials; this allows subsequent evacuation of the vials in the freeze-dryer.
7. Place the vials into metal semicircular trays, then transfer these to a –30°C freezer and incubate for 2 h. The stoppering unit is also precooled to –30°C thus ensuring that when the trays of vials are transferred to the unit, the contents of the vials will not thaw.
8. Switch on the condenser unit of the freeze-drying machine and close the condenser drain valve. When the condenser temperature falls to below –50°C, quickly transfer the stoppering unit and vials to the freeze-drying chamber.

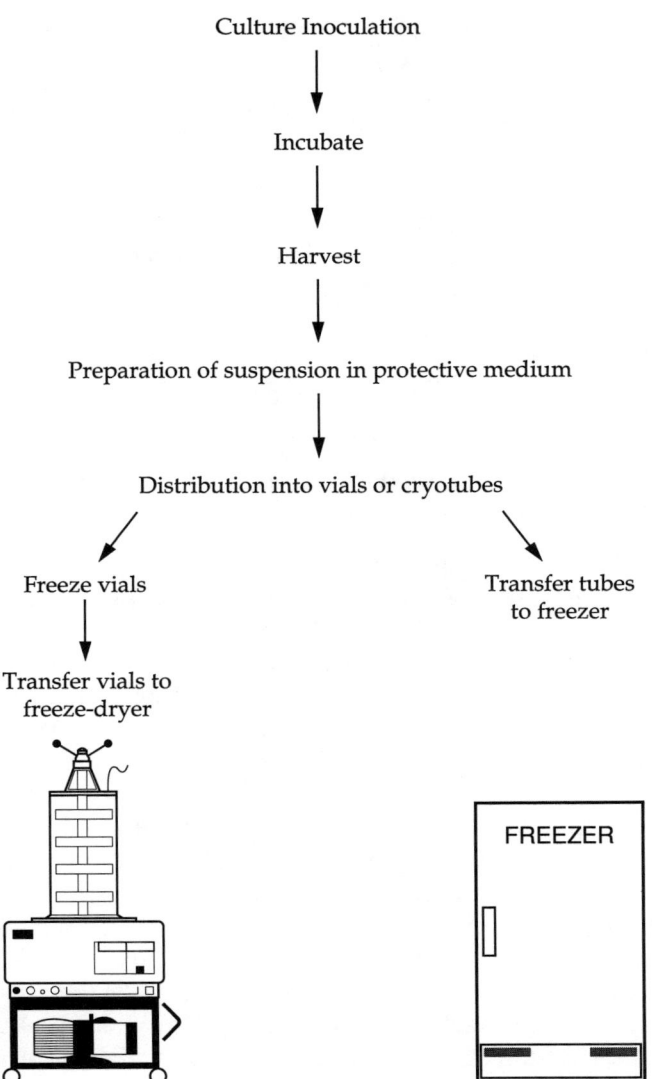

Fig. 1. Flowchart showing the stages involved in the preparation of cultures prior to preservation.

Then apply the vacuum and close the air admittance valve. It is not necessary to use a manifold for secondary drying, as the complete drying program is carried out continuously without further manipulation. After 20 min, check that the chamber is evacuated to a pressure of 300 Pa. The

freeze-dryer should be left to run for 1–24 h to ensure adequate drying of the samples.
9. On completion of the process, stopper the vials under vacuum by rotating the screw jack on the stoppering device (*see* Note 6)
10. Use a high-frequency spark tester (Edwards High Vacuum) to check that a vacuum is maintained within the vials after drying and on storage. A pale blue/violet glow within the vial indicates a satisfactory vacuum, a poor vacuum is indicated by a deep purple glow or no discharge at all.
11. Use aluminium tear off caps to seal the bung on to the vial. For small numbers of vials a hand crimper can be used, but larger numbers of vials are more conveniently sealed with an automatic machine.
12. To recover the bacteria: Taking appropriate precautions (*see* Note 7), remove the metal "tear off" cap using a pair of blunt-nosed forceps. Twist the flap on the upper side of the aluminium cap in the direction of the arrow and remove the cap. Then carefully remove the bung.
13. Using a Pastet, add approx 0.5 mL of nutrient broth to the vial and mix the contents thoroughly (*see* Note 8). Then inoculate the suspension into broth medium and incubate under appropriate conditions (*see* Notes 9–11).
14. Transfer a small aliquot and inoculate (0.1 mL) onto a nutrient agar plate to detect any possible contaminants introduced during opening, or at any other stage of the preservation process.

3.2. Cryopreservation

Figure 1 depicts a flowchart showing the stages involved in the preparation of cultures prior to freezing.

1. Grow the bacteria on an appropriate nonselective, solid medium such as nutrient or blood agar under the optimum growth conditions (*see* Notes 2 and 3).
2. Aseptically add 2 mL of the suspending fluid to the agar slope.
3. Suspend the culture using a sterile plastic loop and thoroughly mix the resulting cell suspension (10^8–10^{10} cells/mL) (*see* Note 4).
4. Aseptically dispense the cell suspension using a sterile plastic Pastet into a cryotube containing presterilized glass beads (*see* Section 2.2., item 3). Aspirate the suspension several times to ensure that air bubbles inside the beads are displaced by the bacterial suspension (*see* Notes 12 and 13).
5. Remove excess liquid from the cryotube leaving the inoculated beads as free of liquid as possible (*see* Note 14).
6. Place the cryovials containing the bacteria/bead mixture in racks of convenient size (Jencons Scientific Ltd., Leighton, Buzzard, Bedfordshire, UK). Then place the racks directly in a commercial freezer that is capable of maintaining temperatures of $-70°C$ (*see* Notes 15 and 16).

7. To recover the bacteria: Remove a cryotube from the freezer and open under aseptic conditions. Using a sterile needle or forceps, remove one bead from the tube. Immediately return the tube to the freezer to prevent the remaining contents from thawing (*see* Note 17).
8. Directly streak the inoculated bead onto a suitable solid medium or alternatively inoculate it into liquid medium. Incubate the inoculated medium under appropriate conditions (*see* Note 18).

4. Notes

1. Experience has shown that the microbiological contamination of the unused bungs is low, and the heat treatment described has proven to be successful in further reducing this to undetectable levels.
2. The use of sloped media in screw-topped 20-mL bottles will reduce the risk of contamination.
3. Depending on the type of organism, the optimal stage of growth for harvesting will vary, but late log phase cultures generally prove suitable for preservation.
4. Harvesting and transfer of cells, particularly of pathogens, should be performed with care to avoid creating aerosols.
5. Ampule identification labels should include a 10-mm gap to the left of the number, with this end placed toward the bottom of the vial. This prevents the dried material from obscuring the identification number and helps to ensure that the figures can always be read from the same end. Alternatively, vials can be labeled on the outside using a labeling machine to print the identification number and organism details.
6. Always ensure that when loading the shelf stoppering unit with trays containing vials, both sides of the unit are balanced, i.e., an even number of trays is used. Failure to do this will mean that the shelves will tilt when screwing down the screw jack unit, and vials will not be completely stoppered.
7. Because the vials are stoppered under vacuum, it is possible that when removing bungs, loose dried material may be released into the atmosphere as an aerosol. Therefore, to minimize the possibility of a laboratory-acquired infection, it is advisable that vials containing pathogens are opened in a safety cabinet.
8. Organisms that are very sensitive to the freeze-drying process may benefit from being left to rehydrate for a few min before subculturing to allow the initiation of repair mechanisms.
9. Physical factors affecting the outcome of freeze-drying have been discussed by Meryman *(3)*. Many of these factors, e.g., incubation temperature of the culture and phase of growth, can be controlled and adjusted to give optimum survival rates. A standard method will work for most species of bacteria, but more fastidious organisms, for instance, anaerobic

photobacteria *(14)*, may require modifications to the protective agents or freeze-drying protocol.
10. Although not necessary for routine use, when investigating problems with loss of viability, it may be useful to measure the moisture content of the finished product. The optimum residual moisture content for survival during storage is about 1% for cells *in vacuo (15)*. Traditional methods of measuring moisture content, such as gravimetric and infrared moisture absorption, are limited in sensitivity and reproducibility. Moisture levels can be accurately and reliably measured by techniques based on the coulometric Karl Fischer method of titration using commercially available equipment (Mitsubishi moisture meter model CA-061, Anachem, Luton, Bedfordshire, UK).
11. In general, the stability of characters is good, but some selection of cells may occur during freeze-drying through loss of viability, which may approach 1000-fold with more delicate organisms. Selection may be increased if serial batches of vials are prepared, i.e., batch four from batch three and batch three from batch two. Cumulative effects from serial selection can be avoided by reserving sufficient vials from batch one to act as a seed stock for subsequent batches.
12. Several different colors of beads are available, allowing for the color coding of different groups of bacteria.
13. Convenient, ready-to-use systems using glass beads are commercially available (Microbank Pro-Lab Diagnostics, Bromborough, Merseyside, UK).
14. Excess suspension left in the tube makes it more difficult to remove individual beads when required.
15. To make the location and retrieval of cultures easier, a record of the contents of each tray and the position of each tray in the freezer should be maintained in a culture collection book.
16. The method described uses a storage temperature of –70°C, but it is now generally accepted that, for long-term maintenance, temperatures below –139°C are preferable. The glass bead technique can, however, also be used for liquid nitrogen storage. Cryotubes can either be transferred from the –70°C freezer to nitrogen, or directly to a dewar containing liquid nitrogen.
17. To prevent rapid thawing of the contents of cryotubes, the use of a well-insulated cryoblock is recommended for transfer of the tubes from the freezer to the bench and back to the freezer. This will enable the operator to remove a number of cryotubes from the freezer for subculture without risk of culture beads defrosting.
18. It has been demonstrated that most cell types have an optimal cooling rate for survival *(16)*, but many species of bacteria will not need to be frozen at

a carefully controlled cooling rate if the suspensions are protected by glycerol. However, if on subculture a high loss of viability is demonstrated, then the use of controlled rate cooling equipment should be considered as an aid to minimizing damage during freezing.

References

1. Kirsop, B. (1985) The current status of culture collections and their contribution to biotechnology. *Crit. Rev. Biotechnol.* **2,** 287–314.
2. Franks, F. (1990) Freeze-drying: from empiricism to predictability. *Cryo-Lett.* **11,** 93–110.
3. Meryman, H. T. (1966) Freeze-drying, in *Cryobiology* (Meryman, H. T., ed.), Academic, London, pp. 609–663.
4. Mackey, B. M. (1984) Lethal and sublethal effects of refrigeration, freezing and freeze-drying on micro-organisms, in *The Revival of Injured Microbes* (Andrew, M. E. E. and Russell, A. D., eds.), Academic, London, Society of Applied Bacteriology Symposium series, No. 12.
5. Crowe, J. H., Carpenter, J. F., Crowe, L. M., and Anchordoguy, T. J. (1990) Are freezing and dehydration similar stress vectors? A comparison of modes of interaction of stabilizing solutes with biomolecules. *Cryobiology* **27,** 219–231.
6. Kirsop, B. E. and Doyle, A. (eds.) (1991) *Maintenance of Micro-Organisms and Cultured Cells: A Manual of Laboratory Methods,* 2d ed., Academic, London.
7. Rudge, R. H. (1991) Maintenance of bacteria by freeze-drying, in *Maintenance of Micro-Organisms and Cultured Cells: A Manual of Laboratory Methods*, 2d ed. (Kirsop, B. E. and Doyle, A., eds.), Academic, London, pp. 31–44.
8. Heckly, R. J. (1978) Preservation of micro-organisms. *Adv. Appl. Microb.* **3,** 1–76.
9. Morris, G. J., Coulson, G. E., and Clarke, K. J. (1988) Freezing injury in *Saccharomyces cerevisiae*: The effect of growth conditions. *Cryobiology* **25,** 471–482.
10. Grout, B. W. W. and Morris, G. J. (1987) Freezing and cellular organization, in *The Effects of Low Temperatures on Biological Systems* (Grout, B. W. W. and Morris, G. J., eds.), Edward Arnold, London, pp. 147–174.
11. Feltham, R. K. A., Power, A. K., Pell, P. A., and Sneath, P. H. A. (1978) A simple method for storage of bacteria at –76°C. *J. Appl. Bact.* **44,** 313–316.
12. Redway, K. F. and Lapage, S. P. (1974) Effect of carbohydrates and related compounds on the long-term preservation of freeze-dried bacteria. *Cryobiology* **11,** 73–79.
13. Barnes, E. M. (1969) Methods for the gram-negative non-sporing anaerobes, in *Methods in Microbiology*, vol. 3B (Norris, J. R. and Ribbons, D. W., eds.), Academic, New York, pp. 151–160.
14. Malik, K. A. (1990) Use of activated charcoal for the preservation of anaerobic phototropic and other sensitive bacteria by freeze-drying. *J. Microbiol. Methods* **12,** 117–124.
15. Nei, T., Souzu, H., and Araki, T. (1966) Effect of residual moisture content on the survival of freeze-dried bacteria during storage under various conditions. *Cryobiology* **2,** 276–279.
16. Mazur, P. (1977) The role of intracellular freezing in the death of cells cooled at supraoptimal rates. *Cryobiology* **14,** 251–272.

CHAPTER 4

Freeze-Drying of Yeasts

Sugio Kawamura, Yukie Murakami, Yukie Miyamoto, and Kazuo Kimura

1. Introduction

Increased interest in biotechnology along with the requirement for strains associated with patent applications to be deposited in a recognized collection, under the terms of the Budapest Treaty (1977), has increased the demand for the maintenance of microorganisms in designated culture collections. As with other microorganisms, long-term preservation methods are required that ensure survival of the strain and the retention of any characteristics of commercial value.

Freeze-drying, which is commonly employed to preserve a variety of microorganisms (*see* Chapters 3 and 6), is also widely used to preserve yeast cultures *(1–3)*. It depends on the removal of water by sublimation of a frozen culture sample under vacuum. The yeast culture suspended in medium containing lyoprotectant is frozen and exposed to a vacuum. Water vapor is trapped and the dried material stored either in an inert gas or under vacuum. Using this technique, a wide variety of strains may be successfully lyophilized and maintained in a preserved state for up to 20 yr. Some yeast strains, including pseudomycelium-forming cultures *(4)*, cannot be preserved by freeze-drying, and alternative preservation methods may be employed (*see* Chapter 5).

The method outlined is one that is routinely used to preserve the majority of yeast strains retained at The National Institute of Bioscience and Human-Technology, Japan (formerly, The Fermentation Research Institute).

2. Materials

1. Growth medium: YM agar medium (*see* Notes 1 and 2). Dissolve 3 g yeast extract, 3 g malt extract, 5 g peptone, 10 g glucose, and 20 g agar in 1 L of distilled water (pH not adjusted). Pour 7 mL of YM agar medium (*see* Note 3) into a test tube, plug with cotton wool, autoclave at 120°C for 15 min, then cool to form an agar slant.
2. Ampules: Tubular glass ampule (*see* Notes 4–6) (8-mm outer diameter, 15-cm long) (*see* Fig. 1). After washing in detergent, rinse in distilled water and dry. Then plug the ampules with cotton wool and place in a metal box. Then heat-sterilize in an oven at 160°C for 2 h before use. Label the ampule with the yeast name and date of lyophilization (*see* Notes 7 and 8).
3. Suspending medium (*see* Note 9): Add 10 g skimmed milk and 1 g sodium glutamate to 100 mL of distilled water. Dispense the resulting mixture in 5–7-mL aliquots into test tubes, plug with cotton wool, and autoclave at 115°C for 15 min. Autoclaving at a higher temperature (i.e., >115°C) causes caramelization of the skimmed milk.
4. Rehydration fluid: 0.9% (w/v) NaCl saline solution. Alternatively, growth medium such as YM medium can be used *(3)*.
5. Freeze-dryer: Manifold type freeze-dryer (*see* Note 10), with cold bath (for prefreezing the sample). The freeeze-dryer should maintain a vacuum of $< 10^{-3}$ torr and the moisture trap and cold bath temperature should be in the range of –30 to –50°C prior to commencing sample freeeze-drying. If the machine has no cold bath, any coolant with a temperature of – 40 to –50°C can be used (*see* Note 11).
6. Sterilized filter paper: Wrap filter papers with aluminum foil or place them in a glass Petri dish and sterilize in an oven at 160°C for 2 h. Alternatively, gauze or cotton wool treated with 70% (v/v) alcohol can also be used.
7. Syringe (1–2.5 mL) with long (18 cm) 18 G needle or Pasteur pipet. Syringe and needle are sterilized before use.
8. Burner: A dual tip or a triple tip O_2/gas burner.
9. Minor items: Test tube (e.g., 16.5 × 165 mm), cotton-wool plug, 70% (v/v) ethanol, gauze, and cotton wool, metal box (for sterilizing ampules in), sterilized inoculation loop and sterilized inoculation probe, ampule cutter or file, high-frequency spark tester, and distilled water.

3. Method

1. All manipulations of cells should be carried out aseptically, in a laminar-flow cabinet or a clean bench. Inoculate the cells onto a YM agar slant medium (*see* Note 12) and incubate under the optimum growth temperature until the culture reaches the end of the logarithmic phase or the beginning of stationary phase (*see* Note 13).

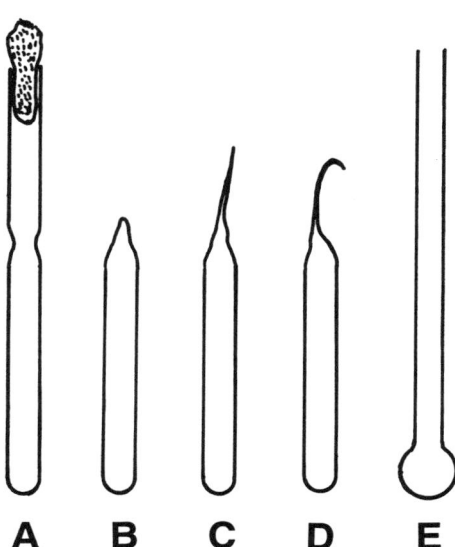

Fig. 1. Manifold freeze-drying ampules. (**A**) Tubular type with cotton plug; (**B**) tubular type, sealed; (**C**) tubular type, sealed (long sharp-pointed); (**D**) tubular type, sealed (long sharp-pointed); (**E**) bulb type. Sealing with long sharp point should be avoided.

2. Check that the vacuum in the freeeze-dryer is <10^{-3} torr, and the moisture trap and cold bath are at the required temperatures (*see* Section 2., item 5 and details in the manual of the machine to be used).
3. Preparation of the cell suspension: Add an aliquot of sterile suspending medium to an agar slant culture (*see* Note 14). Carefully scrape the cells from the agar slant using a sterile inoculation loop to form a uniform cell suspension. Transfer the cell suspension obtained back to the remaining suspending medium in the test tube. Then carefully mix the culture by pipeting in and out using a sterile syringe with long needle or a sterile long Pasteur pipet several times so as to mix thoroughly without bubbling. Cell densities of more than 10^6/mL are commonly used (*3*).
4. Aseptically dispense the cell suspension in 0.2-mL aliquots into sterile ampules using a sterile syringe with long needle or a sterile long Pasteur pipet as in step 3 (*see* Note 15), avoid touching the inside or top of the ampules (*see* Note 16). Plug the ampules with sterile cotton wool.
5. Directly immerse the ampules in a cold bath and quickly stir for ≈ 10 s to ensure that cells are spread on the inner surface of ampules (*see* Notes 17–19). Leave the ampules standing in the cold bath to freeze for more than 2 min.

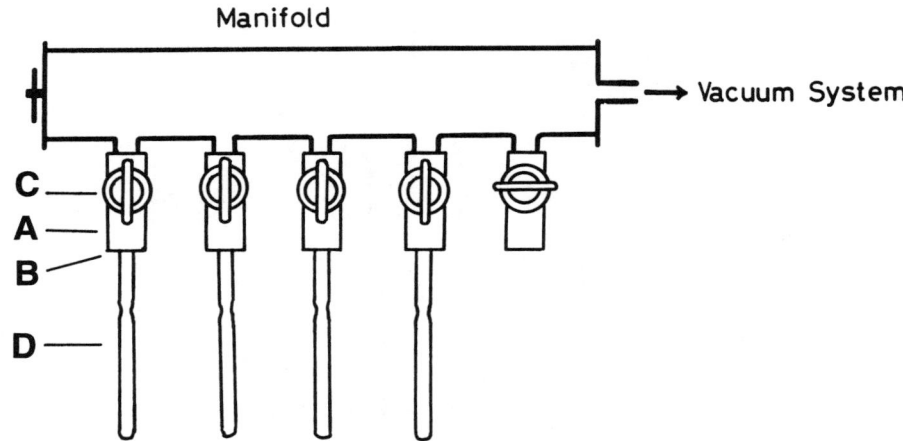

Fig. 2. Manifold freeze-dryer. (**A**) adapter, (**B**) mouth, (**C**) valve, (**D**) tubular type ampule.

6. Rapidly transfer the ampule from the cold bath and disinfect the mouth of adapter where an ampule is inserted and the neck of frozen ampule with 70% (v/v) alcohol. Then remove the cotton wool plug from the neck of the ampule (*see* Note 20) and insert the ampule neck into the manifold adapter. Open the valve to connect the ampule to the vacuum in order to initiate the drying (*see* Note 21). The sample is dried for 4–5 h (*see* Fig. 2).
7. Sealing: On completion of the drying, uniformly heat the narrow part of ampule with a burner. To protect the ampule tip during handling and storage, it should be melted until round rather than long and sharp-pointed (*see* Fig. 1C,D). Gently heat the sealed part of the ampule to prevent cracking during cooling.
8. Touch the neck (sealed area) of the ampule with a high-frequency spark tester to check the degree of vacuum in the ampule. An electric discharge of blue-white to purple in color indicates a good vacuum, and that sealing is successful.
9. Detach the ampule from the manifold and store at 4–5°C in the dark to prevent any light-induced damage.
10. Rehydration and recovery: When required after storage, disinfect the ampule by wiping with 70% (v/v) alcohol. Then score the neck of the ampule with an ampule cutter or a file. Wrap the scored area in sterile filter paper or with sterile cotton wool to prevent the freeze-dried culture from infection or release of the cells on breaking the ampule. Break the ampule

at the scored area and open. Aseptically add 0.2–0.5 mL of rehydration fluid to the freeze-dried sample using a sterile long Pasteur pipet or a sterile syringe with long needle. Gently aspirate several times to thoroughly suspend. Then inoculate the cell suspension into an appropriate growth medium and incubate under optimal conditions.

4. Notes

1. It is easier to harvest cells from solid agar medium; however, higher viabilities may be attained using cultures grown in shake-flask liquid culture *(5)*. For liquid culture, prepare 20 mL of medium (without agar) in a 100-mL flask capped with a cotton wool bung, sterilize as detailed in Section 2., item 1.
2. Minimum growth medium is sometimes used for recombinant cells to avoid the risk of missing plasmids *(6)*, although complete growth medium (e.g., YEPD agar medium) is used when growth in the minimum medium is not sufficient. Minimum growth medium: Yeast nitrogen base without amino acid (Difco, Detroit, MI), 0.67% (w/v) glucose, 2% (w/v) amino acids. YEPD agar medium: Add 20 g peptone, 20 g glucose, 10 g yeast extract, and 20 g agar to 1 L of distilled water.
3. Four agar slants are prepared for storage of every yeast strain. Two are used for culture and the other two for recovery. One slant produces sufficient cell suspension for 20 ampules.
4. Poor quality glass may result in cracks in the ampule during long-term preservation. Pyrex or Borosilicate glass ampule should be used. Neutral soft glass ampules are sometimes used, especially for pathogenic yeast, in order to prevent dispersion of dried cells on breaking the ampule *(1)*.
5. Specially manufactured ampules with narrow diameter in the middle should be used.
6. Tubular type ampules are used for freeze-drying samples using skimmed milk as lyoprotectant, and bulb type ampules when sucrose suspending media are employed.
7. Ampules may be labeled using a labeling machine (Markem Labeling Machine, Markem Corporation, USA). After labeling, ampules are dry sterilized at 160°C for 10 min in an oven.
8. On using labels that are inserted inside the ampule, ENM nontoxic quick drying ink (Rexel) should be used *(3)*. The size of label depends on the size of ampule (e.g., 5 × 30 mm) *(1)*. An ampule with label is autoclaved and dried in an oven, instead of sterilizing in an oven *(1)*.
9. Alternatively serum *(6,7)* or serum-based suspending medium *(3,5,8)* may be used. Recently, either the skimmed milk and sodium glutamate or the

sucrose-based suspending media have been most commonly used *(3,4,9–11)*. Using these mixtures generally results in high viability levels after freeze-drying. Some of the most widely used suspending media are detailed: 7.5% (w/v) glucose in inactivated horse serum *(3,5)*; 10% (w/v) glucose and 10% (w/v) sodium glutamate *(12)*; 5% (v/v) dextran solution containing 1% (w/v) sodium glutamate *(13)*; 20% (w/v) skimmed milk and 10% (w/v) sodium glutamate *(3)*; 10% (w/v) skimmed milk and 1% (w/v) sodium glutamate *(9)*; 3% (w/v) skimmed milk, 1% (w/v) sodium glutamate, and 5% (w/v) sucrose *(4,10)*; 10% (w/v) skimmed milk, 1% (w/v) sodium glutamate, and 10% (w/v) sucrose *(11)*.
10. Alternative format freeze-dryers include chamber type *(2)* and centrifuged type *(3)*.
11. Alternative coolants include: crushed dry ice, ethyl cellosolve (–30 to –50°C) *(2)*; crushed dry ice, 95% (v/v) ethanol (min –75°C). Liquid nitrogen is generally not used, as the resultant high rate of cooling sometimes causes death of the yeast cells.
12. Alternatively cultivate in liquid medium. Add 1 mL of inoculum to 20 mL of medium and incubate on an orbital shaker at 37°C.
13. The age of the culture is important. Cultures harvested at the end of logarithmic phase or beginning of stationary phase survive freeze-drying better than those harvested at other phases of growth *(4)*.
14. Liquid culture: The culture is aseptically harvested by centrifugation at 5000*g* for 10 min. Decant the supernatant. Resuspend the cells in an appropriate volume of sterile suspending medium with a sterile inoculation probe. Pour the cell suspension back into the rest of suspending medium and mix thoroughly. The remaining procedure is the same as the case of the agar slant culture.
15. If viability estimation counts are required, use an ampule without freeze-drying as a control; this is kept in an ice-bath until viability counts are performed. Spread 0.1-mL aliquots of logarithmic dilutions of the cell suspension on appropriate medium and incubate under standard conditions.
16. When cells stick on the inside or top of the ampule, they carbonize on sealing the ampule.
17. Sample should be frozen, if possible, within a few minutes, or at least within 1 h. Standing for a long time results in settling of cells *(1)* and a decrease of viability *(5)*.
18. Usually 3–4 ampules can be handled at a time.
19. On using bulb type ampules, shell freezing is performed; that is, the ampule is rotated to spread cells on the inner surface of the ampule.
20. Cotton wool plugs are sometimes not removed, especially in the case of pathogenic yeast. Freeze-drying with cotton wool plug in place will reduce

the risk of sample contamination and protect the laboratory personnel from infection. Cotton wool plugs are pushed down in the ampule with a sterile inoculation probe, and flamed to remove any residual cotton fiber on the top. Residual cotton may prevent an adequate seal being made and result in a loss of vacuum within the stored ampule.
21. If samples melt during drying, the procedure has failed; this may be caused by an inadequate level of vacuum. The entire procedure should be repeated from step 1.

References

1. Lapage, S. P., Shelton, J. E., Mitschell, T. G., and Mackenzie, A. R. (1970) Culture collections and the preservation of bacteria, in *Methods in Microbiology*, vol. 3A (Norris, J. R. and Ribbons, D. W., eds.), Academic, London, pp. 135–228.
2. Alexander, M., Daggatt, P.-M., Gherna, R., Jong, S., and Simione, F. Jr. (1980) *American Type Culture Collection Methods 1. Laboratory Manual on Preservation Freezing and Freeze-Drying: As Applied to Algae, Bacteria, Fungi and Protozoa* (Hatt, H., ed.), American Type Culture Collection, Rockville, MD, pp. 3–45.
3. Kirsop, B. E. (1984) Maintenance of yeasts, in *Maintenance of Microorganisms: A Manual of Laboratory Methods* (Kirsop, B. E. and Snell, J. J. S., eds.), Academic, London, pp. 109–130.
4. Tsubouchi, J. and Takada, N. (1974) *Sporobolomyces odorus* no touketsu narabini kansou. *Jpn. J. Freezing Drying* **20**, 24–28. (in Japanese)
5. Rose, D. (1970) Some factors influencing the survival of freeze dried yeast cultures. *J. Appl. Bact.* **33**, 228–232.
6. Botstein, D., Falco, S. C., Stewart, S. E., Brennan, M., Scherer, S., Stinchcomb, D. T., Struhl, K., and Davis, R. W. (1979) Sterile host yeasts (SHY): a eukaryotic system of biological containment for recombinant DNA experiments. *Gene* **8**, 17–24.
7. Benedict, R. G., Sharpe, E. S., Corman, J., Meyers, G. B., Baer, E. F., Hall, H. H., and Jackson, R. W. (1961) Preservation of microorganisms by freeze-drying. II. The destructive action of oxygen. Additional stabilizers for *Serratia marcescens*. Experiments with other microorganisms. *Appl. Microbiol.* **9**, 256–262.
8. Lapage, S. P., Shelton, J. E., and Mitschell, T. G. (1970) Media for the maintenance and preservation of bacteria, in *Methods in Microbiology*, vol. 3A (Norris, J. R. and Ribbons, D. W., eds.), Academic, London, pp. 1–133.
9. Abe, S. (1971) Raio-fairu niyoru biseibutu hozon. *Jpn. J. Freezing Drying* **17**, 5–14. (in Japanese)
10. Takada, N., Takano, M., and Terui, G. (1972) *Brettanomyces* oyobi *Sporobolomyces* no touketu kansou. *Jpn. J. Freezing Drying* **18**, 1–5. (in Japanese)
11. Kimura, K., Aikawa, T., and Ito, J. (1978) Studies on preservation of industrially useful microorganisms by freeze-drying techniques. On yeast cells. *Jpn. J. Freezing Drying* **24**, 26–30. (in Japanese)
12. Hall, J. F. and Webb, T. B. J. (1975) Factors affecting the survival of lyophilized brewery yeast strains. *J. Inst. Brew.* **81**, 471–475.
13. Atteunisse, J. (1972) Viability of lyophilized microorganisms after storage. *Antonie van Leeuwenhoek* **39**, 243–248.

CHAPTER 5

Cryopreservation of Yeast Cultures

Chris J. Bond

1. Introduction

Yeast cultures are held in long-term storage in the National Collection of Yeast Cultures (NCYC; Norwich, UK) by two methods: freeze-dried in glass ampules and under liquid nitrogen using glycerol as a cryoprotectant. Freeze-drying is a generally accepted method for yeast storage, having the advantages of conferring longevity and genetic stability, as well as being suitable for easy worldwide postal distribution of the cultures in glass ampules. However, preservation by freeze-drying tends to be much more labor intensive than storage in liquid nitrogen and requires a higher level of skill to produce an acceptable product. Strain viabilities are generally low, typically being between 1 and 30%, as compared to >30% for those of yeast preserved frozen in liquid nitrogen. There are also several yeast genera, including *Lipomyces, Leucosporidium,* and *Rhodosporidium* which have particularly low survival levels and frequently cannot be successfully freeze-dried by the standard method. However, some improvements have been made recently using trehalose as a protectant *(1,2)*. Techniques for freeze-drying yeasts can be found in Chapter 4.

Storage of cultures in liquid nitrogen, although technically simple, can involve relatively high running costs because of the necessity of regular filling of the containers. The initial cost of the equipment is comparable to that used for freeze-dying but the costs and problems associated with the handling of liquid nitrogen have led some collections to seek alternatives *(3)*. However, for most workers the technique of liquid nitrogen storage is convenient, well tried and tested, and unlikely to be superseded in the near future.

The method presented here uses heat-sealed straws and is a miniaturized version of the one that is commonly used. This enables a considerable reduction in storage space and extra protection against contamination by liquid nitrogen leakage. Storage in straws was first described in 1978 by Gilmour et al. *(4)* using artificial insemination straws, and variations on the original method are now in use in a large number of laboratories around the world. Work on refining methods of storage in liquid nitrogen is continuing, as is research into the effects of the freezing process on the cells. The following paragraphs give an outline of the current understanding.

During the process of liquid nitrogen storage, certain changes take place in the cells and their immediate environment *(5)*. As the straws are cooled, extracellular ice formation results in an increase in the solute concentration around the cells causing them to lose water and shrink *(6,7)*. This freeze-induced dehydration causes the cell wall to decrease in surface area and increase in thickness. As the maximum packing density of the lipids in the cell membrane bilayer is reached, its normal structure changes. Membrane invaginations occur to allow the cells to shrink further as water is removed. This process is reversible, provided that none of the membrane material becomes lost within the cytoplasm, and on thawing the cell will return to its normal volume.

Cell shrinkage during freezing is vital to prevent cellular damage, hence the need to select the correct cooling rate. If the cooling rate is too rapid there is insufficient time for the cell to lose water and intracellular ice formation then occurs, which causes damage to cell organelles *(7)*. Genetic damage may occur if the nucleus becomes disrupted and plasmids may also be destroyed. Mitochondrial damage may also occur and will result in respiratorily deficient cells, giving rise to petite colonies.

2. Material

The procedures described involve use of potentially hazardous materials. The relevant local safety regulations (e.g., Control of Substances Hazardous to Health [COSHH] regulations *[8]*) should be consulted prior to implementation of these procedures.

1. Difco yeast malt (YM) broth: Difco dehydrated YM broth (Ref. no. 0711-01, Difco Inc., Detroit, MI), 21 g/L. Alternatively, use YM medium: 3 g yeast extract, 3 g malt extract, 5 g peptone, and 1 g glucose, made up to 1 L.

After mixing, the pH should be between 5 and 6. Dispense in 10-mL amounts into suitable bottles. Sterilize by autoclaving for 15 min at 121°C.
2. YM agar: Add 2% (w/v) agar to Difco YM broth or YM media before sterilization. After mixing, the pH should be between 5 and 6. Sterilize by autoclaving for 15 min at 15-lb pressure (121°C). Dispense in 20-mL aliquots into sterile Petri dishes and leave to cool.
3. 10% (v/v) Glycerol cryoprotectant solution: Dissolve 10 mL glycerol in 90 mL distilled water. Sterilize by filtration through a 0.22-μm filter and dispense in 15-mL aliquots into suitable bottles. Store at room temperature.
4. Straws: Cut colored polypropylene drinking straws (order no. SJ012, 4-mm diameter, Key Catering Ltd., London, UK) to 2.5-cm lengths (*see* Note 1). Seal one end of each straw by holding firmly with nonridged forceps 2 mm from the end to be sealed and bringing the projecting end to 1 cm from the flame of a Bunsen burner (*see* Fig. 1; Note 2). Place the straws in a glass Petri dish and sterilize by autoclaving at 121°C for 15 min. For ease of handling, the straws should be evenly spaced around the edge of the dish with all the open ends pointing in the same direction. Two long unsealed straws should also be prepared for use as rests for straws awaiting final sealing (*see* Fig. 2; Note 3). These should also be placed in glass Petri dishes and sterilized. Ensure that the straws are dry before use, using a moderate temperature (40–60°C) drying cabinet if necessary.
5. Nunc cryotubes: Plastic screwcap 1.8-mL ampules are available sterilized from the manufacturer (Life Technologies Ltd., Inchinnan, UK).
6. A refrigerated methanol bath precooled to –30°C (*see* Note 4).
7. Liquid nitrogen containers: Cryogenic storage containers with liquid phase storage racks and dividers to store 2-mL cryotubes (Jencons Scientific Ltd., Leighton, Buzzard, UK) (*see* Fig. 3).
8. Safety equipment: Cryogloves, goggles, and so on.

3. Method

Follow good microbiological practice and aseptic techniques throughout.

1. Grow the culture to be frozen in 10 mL of YM broth for 72 h at 25°C on a reciprocal shaker.
2. Mix equal amounts of inoculum and glycerol cryoprotectant solution aseptically in a sterile bottle (*see* Note 5). Use forceps, sterilized by immersion in alcohol followed by flaming, to remove a single straw from the Petri dish. Gently grip the straw about halfway along its length to allow the insertion of the end of a Pasteur pipet containing the inoculum. Insert the pipet until the pipet tip is at the sealed end of the straw and then withdraw

Fig. 1. Sealing polypropylene straws.

it as the inoculum fills the straw. Fill the straw to approx two-thirds of its capacity. On withdrawing the pipet from the straw, any excess inoculum can be sucked back into the pipet (*see* Notes 6 and 7).
3. The open end of the straw is sealed as described in Section 2., item 4.
4. Test the straws for leaks by holding them with forceps above the surface of a suitable disinfectant in a high-sided beaker and then squeezing them. Any liquid forced out of the seals will be safely contained within the beaker. Discard leaking straws and autoclave them.
5. Place six straws in each 1.8-mL cryotube. (If postthaw cell viability counts are required, a single straw may be placed in a separate ampule for ease of recovery.) Mark each straw and cryotube with the relevant strain designation and date of freezing using a black Pentel permanent marker pen.
6. Place the filled cryotubes in the methanol bath, which has been precooled to −30°C, for 2 h to allow dehydration (*see* Note 8).
7. Transfer the cooled cryotubes to the liquid nitrogen containers and place in the racking (inventory) system (*see* Notes 9 and 10). Note the position of

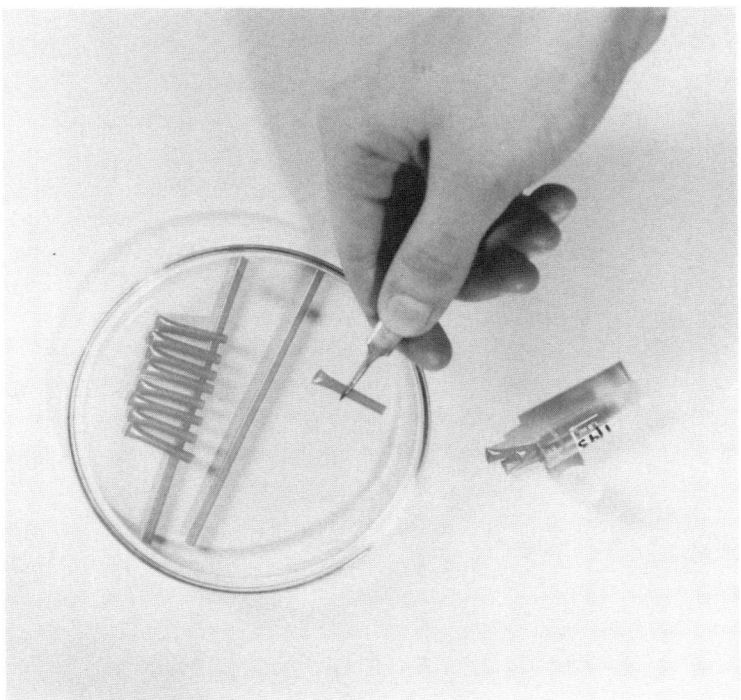

Fig. 2. Filled polypropylene straws sealed at one end awaiting final sealing and (right) sealed straws in cryotube awaiting transfer to methanol bath.

the cryotubes (*see* Note 11). Remove excess methanol from the outside of the cryotubes to prevent it from freezing the tubes to the racking system while immersed in the liquid nitrogen (*see* Notes 12–14).

8. After storage: Locate the cryotube and remove it from the racking system; check the strain number and date of freezing written on the tube. Remove a single straw and replace the cryotube in the racking system (*see* Note 15). Rapidly transfer the straw to a water bath, incubate at 35°C for 30 s, and agitate to ensure rapid and even thawing takes place (*see* Note 16).
9. Remove the straw from the water bath and dry. Grip one end of the straw and sterilize the other end by wiping with alcohol. Cut off the sterile end with scissors that have been flamed with alcohol. Then remove the contents using a Pasteur pipet inserted into the open mouth of the straw. Mix the contents by repeated pipeting before transferring as an inoculum to suitable growth media (*see* Note 17).

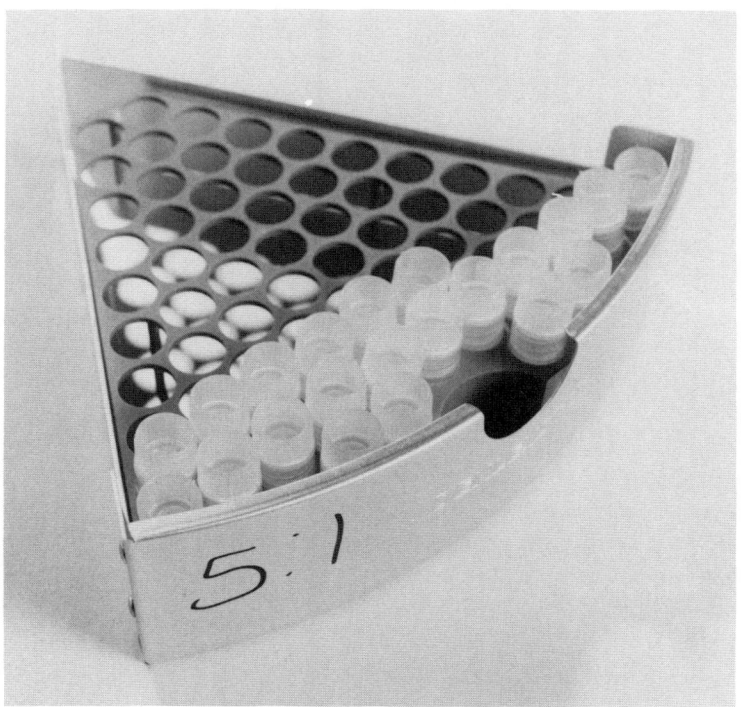

Fig. 3. Filled cryotubes in numbered racking system.

10. Viability counts:
    a. On cultures prior to freezing: Add 1 mL of the original cell suspension to 9 mL of sterile glass-distilled water. Prepare further logarithmic dilutions to 10^{-6}. Transfer three drops from a 30-dropper pipet (0.1 mL) of dilutions 10^{-6} to 10^{-3} onto YM agar. Incubate the plates at 25°C for 72 h, or longer if necessary (*see* Note 18).
    b. On cultures after freezing: Add two drops of the thawed cell suspension (0.06 mL) to 0.54 mL of sterile glass-distilled water. Transfer 0.5 mL of this 1:10 dilution to 4.5 mL of sterile glass-distilled water, prepare additional logarithmic dilutions to 10^{-6} as detailed in step a, then inoculate YM agar plates, incubate, and count as in step a. The percentage viability of the culture is calculated and recorded.

4. Notes

1. The NCYC uses drinking straws in preference to insemination straws due to their low cost, robust nature, and ability to withstand autoclaving.

2. Polypropylene straws should melt quickly and will form a strong seal that will set firm within 3–4 s of being removed from the heat. Care should be taken not to burn the polypropylene or deform the rest of the straw. A standard Bunsen burner may be used, although a "fishtail" will give a more controllable flame. Occasionally leaks occur at the corners of the straw; particular attention should be paid to these areas to ensure they are properly sealed. Some workers have used impulse heat sealers to seal polypropylene straws. In our experience these tend to produce weak seals and do not produce the characteristic "lip" formed by the melted polypropylene that is useful when removing straws from cryotubes.
3. If several straws are being filled with the same inoculum it is convenient to transfer each straw to a sterile Petri dish before it is sealed. The straws should be leaned with the open end against a second longer straw to allow their easy removal for sealing and to prevent leaking from the opening (*see* Fig. 2). The straws are then sealed and placed in the cryotube. This is more efficient than filling and sealing each straw separately.
4. **Caution, methanol is toxic and flammable. Avoid skin contact.**
5. Other workers have successfully used cryoprotectants other than glycerol. Some have been used with a wide range of yeasts, others with single strains only. Substances used include 10% (v/v) dimethyl sulfoxide, 10% (v/v) ethanol, 10% (v/v) methanol, 10% (v/v) YM broth, and 5% (w/v) hydroxyethyl starch.
6. Care should be taken not to allow droplets to remain on the open end of the straw as these can prevent proper sealing.
7. The final cell concentration in the straws once the inoculum and cryoprotectant are mixed is between 10^6 and 10^7 cells/mL. Final glycerol concentration in the straws is 5% (v/v).
8. Experiments on two test strains of *Saccharomyces cerevisiae* at the NCYC showed no significant variation in viability for cells frozen to –20, –30, or –40°C for 1, 2, or 3 h during primary freezing. The NCYC uses the intermediate range, which has so far been successful for all NCYC strains. A study carried out by Pearson et al. *(5)* indicated that cooling rates >8°C/min result in a sharp drop in cell viability and cause irreversible genetic damage. Cells have been found to be more sensitive to cooling rates than to glycerol concentration *(5)*.
9. Care should be taken to provide adequate ventilation where liquid nitrogen is in use, as a buildup of nitrogen can cause suffocation. Workers should avoid traveling in elevators when transporting liquid nitrogen containers. Goggles and gloves should be worn while carrying out operations using liquid nitrogen.

10. Several models of liquid nitrogen storage container are available. Some offer greater amounts of storage space by reducing the level of liquid nitrogen above the racking system to a minimum. In order to ensure that the top racks of these containers are always completely submerged, they must be filled more regularly than those with less height of racking. These containers are best avoided unless an automatic "top-up" system is available.
11. Since stock levels cannot be easily checked once immersed in liquid nitrogen, accurate record keeping of both stock levels and the position of each strain in the racking system is important. Computerized stock control systems are ideal for storage of this information. The position of each cryotube in a rack may be mapped and recorded in the following way, for example:

Strain number:	240
LN_2 container:	A
Section:	3
Tray:	2
Row:	1
Position:	6
Total no. of straws:	6

If only a small number of strains are to be stored, different colored caps can be used for color coding of cryotubes.
12. If there is a large distance between the methanol bath and liquid nitrogen containers, a precooled dewar should be used to transport the tubes to prevent warming.
13. Cultures can be stored successfully in the vapor phase of liquid nitrogen. However, since changes may still occur in the stored cells at temperatures above $-139°C$, storage in the liquid phase at $-196°C$ is preferable. While the cultures are submerged in the liquid nitrogen, temperature stability is guaranteed. If storage is only required for short periods, higher temperatures may be adequate.
14. The long-term survival of yeasts held in liquid nitrogen has not been well documented, but all evidence suggests that losses during storage are insignificant. The NCYC has found no drop in viability of cultures stored for up to 10 yr. Work with mutant strains of *S. cerevisiae* and *Schiz. pombe* at the NCYC has also shown that genetic stability is also very good.
15. If the racking system used holds many cryotubes they will all be exposed to higher temperatures when the racking system is removed from the liquid nitrogen to recover a straw. Care should be taken to minimize this time as much as possible. Work done at the NCYC has suggested that using straws sealed inside cryotubes offers considerable protection against short

exposure to higher temperatures. No significant drop in viability has been recorded in straws held in a racking system that has been repeatedly removed briefly from the liquid nitrogen.
16. Nunc recommend that cryotubes should not be used for freezing in liquid nitrogen unless correctly sealed in Cryoflex since trapped nitrogen can expand and cause the tubes to rupture once they are removed from the liquid. Since, in this method, the cryotube is only being used for secondary containment, the screwcap should not be firmly tightened. This will allow trapped nitrogen to leave the tube safely as its temperature increases. However, tubes should always be held at arm's length when being removed and safety goggles and gloves should be worn. Occasionally straws will themselves rupture on thawing because of poor seals letting in liquid nitrogen. Provision should be made for sterilizing bench tops and equipment if they become contaminated with the contents of the ruptured straw.
17. Cells removed from liquid nitrogen storage should be transferred initially to small aliquots (10 mL) of suitable growth media. Larger volumes of media can be inoculated from this culture once sufficient growth has occurred.
18. The plates are kept horizontal at all times to ensure the drops remain discrete. Dilutions containing 20–30 colonies are used for estimating viability. The number of cells/mL inoculated into the straw is equal to the number of colonies in three drops multiplied by 10 times the dilution factor.

References

1. Berny, J. F. and Hennebert, G. L. (1991) Viability and stability of yeast cells and filamentous fungus spores during freeze-drying—effects of protectants and cooling rates. *Mycologia* **83,** 805–815.
2. Roser, B. (1991) Trehalose drying: a novel replacement for freeze-drying. *Biopharm.—Technol. Bus. Biopharm.* **4,** 47–53.
3. Mikata, K. and Banno, I. (1989) Preservation of yeast cultures by L drying: viability after 5 years of storage at 5°C. *IFO Res. Comm.* **14,** 80–103.
4. Gilmour, M. N., Turner, G., Berman, R. G., and Krenzer, A. K. (1978) Compact liquid nitrogen storage system yielding high recoveries of gram-negative anaerobes. *Appl. Environ. Microbiol.* **35,** 84–88.
5. Pearson, B. M., Jackman, P. J. H., Painting, K. A., and Morris, G. J. (1990) Stability of genetically manipulated yeasts under different cryopreservation regimes. *Cryo-Lett.* **11,** 205–210.
6. Diller, R. R. and Knox, J. M. (1983) Automated computer-analysis of cell-size changes during cryomicroscope freezing—a biased trident convolution mask technique. *Cryo-Lett.* **4,** 77–92.
7. Morris, G. J., Coulson G. E., and Clarke K. J. (1988) Freezing injury in *Saccharomyces cerevisiae*—the effect of growth conditions. *Cryobiology* **25,** 471–482.
8. Control of Substances Hazardous to Health Regulations (1988) Approved code of practice, HMSO, London.

Chapter 6

Cryopreservation and Freeze-Drying of Fungi

Jacqueline A. Kolkowski and David Smith

1. Introduction

Many aspects of mycology today necessitate living cultures. Traditional taxonomy is being supplemented by molecular biological techniques often requiring growing organisms. The need to keep particular strains for confirmation of results, further studies, and as references is well established.

The main requirement of a preservation technique is to maintain the fungus in a viable and stable state without morphological, physiological, or genetic change until it is required for further use. Therefore conditions of storage should be selected to minimize the risk of such change. Culture preservation techniques range from continuous growth methods that reduce the rate of metabolism to an ideal situation where metabolism is considered suspended or nearly so *(1,2)*. No single technique has been applied successfully to all fungi, although storage in or above liquid nitrogen is now well developed for many fungal strains.

One factor common to all methods is the need to start with a healthy culture to obtain the best results. Optimal growth conditions including temperature, aeration, humidity, illumination, and media must be found *(1)*.

Growth requirements vary from strain to strain, although cultures of the same species and genera usually grow best on similar media. A medium that induces good sporulation and minimal mycelium formation is desirable for successful freeze-drying *(1)*, although this is not necessary if the culture is to be cryopreserved. In some cases it is often better to grow the organism under osmotic stress as chemicals accumulated in the cytoplasm under these conditions also protect during cooling.

From: *Methods in Molecular Biology, Vol. 38: Cryopreservation and Freeze-Drying Protocols*
Edited by: J. G. Day and M. R. McLellan Copyright © 1995 Humana Press Inc., Totowa, NJ

1.1. Cryopreservation

Lowering the temperature of biological material reduces the rate of metabolism until all internal water is frozen and no further biochemical reactions occur *(3)*. This is quite often lethal and at the very least causes cellular injury. Although fungi are quite often resistant to ice damage, cooling must be controlled to achieve optimum survival. The avoidance of intracellular ice and the reduction of solution or concentration effects are necessary *(4)*. Little metabolic activity takes place below −70°C. However, recrystallization of ice can occur at temperatures above −139°C *(5)*, and this can cause structural damage during storage. Consequently, cooling protocols have to be carefully designed for cells in order to inflict least damage.

The cryopreservation of microfungi at the ultra-low temperature of −196°C in liquid nitrogen or the vapor above, is currently regarded as the best method of preservation *(6,7)*. It can be widely applied to sporulating and nonsporulating cultures. Initial work with fungi was undertaken by Hwang *(8)*, who employed a method designed for freezing avian spermatozoa *(9)*. Similar methods have been used successfully by many workers *(7,10,11)*. Provided adequate care is taken during freezing and thawing, the culture will not undergo any change, either phenotypically or genotypically. Optimization of the technique for individual strains has enabled the preservation of organisms that have previously been recalcitrant to successful freezing *(10)*.

The choice of cryoprotectant is a matter of experience and varies according to the organism. Glycerol gives very satisfactory results but requires time to penetrate the organism; some fungi are damaged by this delay. Dimethyl sulfoxide (DMSO) penetrates rapidly and is often more satisfactory *(12,13)*. Sugars and large molecular substances, such as polyvinyl pyrrolidine (PVP) *(14)*, have been used *(15)* but in general have been less successful.

At the International Mycological Institute (IMI; Egham, UK), over 4000 species belonging to 700 genera have been successfully frozen in 10% (v/v) glycerol. Few morphological or physiological changes have been observed *(16)*.

Finding the optimum cooling rate has been the subject of much research *(4,8,10)*. Slow cooling at 1°C/min over the critical phase has proved most successful *(16,17)*, but some less sensitive isolates respond well to rapid cooling, preferably without protectant. Slow warming may cause damage owing to the recrystallization of ice, therefore rapid thawing is recommended. Slow freezing and rapid thawing generally give high recoveries for fungi *(18)*. Storage at −196°C in the liquid nitrogen or at slightly higher temperatures in the vapor phase is employed at IMI.

1.2. Freeze-Drying (Lyophilization)

Lyophilization (preservation by drying under reduced pressure from the frozen state by sublimation of ice) was first used for fungi by Raper and Alexander *(19)*. Improvements in methods and equipment over the years have led to a reliable and successful preservation technique for sporulating fungi. Stability, long shelf-life, simple storage, and easy distribution are the main advantages of freeze-drying *(18,20,21)*.

A freeze-drying system incorporates freezing the suspension, generating a vacuum; and absorbing the water vapor evolved. The protectant used, rate of cooling, final temperature, rate of heat input during drying, residual moisture, and storage conditions all affect the viability and stability of fungi *(22,23)*.

The suspending medium should give protection to the spores from freezing damage and during storage. Media most often used are skimmed milk, serum, peptone, various sugars, or mixtures of these.

The rate of freezing is a very important factor; this must be optimized to achieve best recovery. Slow freezing rates are employed, 1°C/min is the rate normally quoted *(1,18)*. The technique of evaporative cooling can be used successfully for the storage of many sporulating fungi *(1)*.

2. Materials

2.1. Materials for Liquid Nitrogen Storage

1. Biological safety cabinet and facilities for carrying out microbiological methods and aseptic techniques (*see* Note 1).
2. Cultures should be healthy and exhibit all characters to be preserved, both morphologically and physiologically (*see* Note 2).
3. Medium as required for different fungi *(1)*.
4. Borosilicate glass ampules (1 mL) covered with single aluminum foil lids sterilized at 180°C for 3 h or alternatively presterilized 2.0-mL cryotubes (*see* Note 3).
5. Cryoprotectant: 10% (v/v) glycerol in distilled water dispensed in 10-mL amounts autoclaved 121°C for 15 min.
6. Liquid nitrogen wide-necked storage tank with drawer rack inventory control system.

7. Safety equipment: Should include cryogloves (cold resistant), face shield, forceps, and oxygen monitor.
8. 1% (w/v) Solution of erythrosin B.
9. Air/gas sealing torch.

2.2. General Materials for Freeze-Drying

1. Biological safety cabinet and facilities for carrying out microbiological methods safely (*see* Note 1).
2. Cultures should be healthy and sporulating, exhibiting all characters to be preserved both morphologically and physiologically (*see* Note 4).
3. Growth media as required for different fungi *(1)*.
4. 10% (w/v) Skimmed milk, 5% (w/v) inositol in distilled water dispensed in 10-mL amounts in glass universal bottles. These are autoclaved at 114°C for 10 min (*see* Note 5).
5. High voltage vacuum spark tester.

2.3. Materials for Spin Freeze-Drying

1. Neutral glass ampules (0.5 mL) labeled with the culture number and covered with lint caps to prevent aerial contamination (*see* Note 6). These are heat sterilized at 180°C for 3 h.
2. Metal racks for supporting 0.5-mL neutral glass ampules.
3. Freeze-dryer with spin freeze and manifold drying accessories.
4. Nonabsorbent cotton wool sterilized at 180°C for 3 h.
5. Air/gas constricting torch.
6. Air/gas sealing torch.
7. Heat-resistant mat.
8. Di-phosphorus pentoxide general purpose reagent (GPR) (*see* Note 7).

2.4. Materials for Shelf Freeze-Drying

1. Flat bottomed preconstricted glass (ampules) vials (2 mL) labeled with the culture number. These are covered with aluminum foil to prevent aerial contamination then heat sterilized by autoclaving at 180°C for 3 h.
2. Metal racks for holding 2-mL flat-bottomed glass vials.
3. Grooved rubber bungs sterilized by autoclaving at 121°C for 15 min then placed in 70% (v/v) industrial spirits (*see* Note 8).
4. Shelf freeze-dryer with programmable shelf temperature control.

3. Methods

3.1. Liquid Nitrogen Storage

All culture work should be carried out using aseptic techniques in a microbiology safety cabinet (*see* Note 1).

1. Grow cultures under optimal growth conditions and on suitable media (*see* Section 2.1., items 1, 2, and 3; Notes 2 and 9).
2. Prepare a spore or mycelial suspension in sterile 10% (v/v) glycerol; mechanical damage must be avoided (*see* Note 10).
3. Add 0.5-mL aliquots of suspension to the sterile ampules or cryotubes; label with the culture number using a permanent ink marker.
4. Seal the glass ampules using an air/gas torch, ensuring that the liquid suspension does not come into contact with the hot glass. Rotate the extended neck of the ampule in the flame and when the glass is molten use forceps to support and gently draw the glass so that the flame cuts through, leaving the base of the ampule sealed.
5. Test the seals of the ampules for leakage in a water bath containing erthrocin B dye bath at 4–7°C; this also precools the fungal suspension. Allow at least 1 h for the cells to equilibrate in the glycerol (*see* Note 11). Plastic cryotubes are simpler to handle and require the caps to be screwed down tightly after filling.
6. Place the ampules or cryotubes in racks and cool at approx 1°C/min in a controlled rate cooler (*see* Notes 12 and 13). Alternatively place the ampules/cryotubes in the neck of the nitrogen refrigerator at −35°C for 45 min (*see* Note 12).
7. Transfer the ampules into the liquid nitrogen; this cools them to −196°C, leave for 2 min to complete the cooling, then transfer to the storage racks and store in the vapor phase.
8. Record the location of each culture in the inventory control system.
9. After 3–4 d retrieve an ampule from the refrigerator to test viability and purity of the fungus.
10. Warm the ampule rapidly by immersion in a water bath at +37°C. Remove the ampule immediately on completion of thawing and do not allow to warm up to the temperature of the bath. Alternatively thaw the ampules in a controlled rate cooler on a warming cycle (*see* Note 14).
11. Opening of the ampule or cryotube and the transfer to media should be carried out in a class II microbiological safety cabinet. Surface sterilize the glass ampules by immersion or wiping with 70% (v/v) alcohol. Open the ampules by scoring the neck and snapping off the top of the tube. Aseptically transfer the contents to a suitable growth medium. Alternatively remove the plastic screwcap of the cryotube and aseptically remove the contents using a Pasteur pipet and transfer on to suitable growth medium.

3.2. Spin Freeze-Drying

All culture work should be carried out in an appropriate microbiological safety cabinet (*see* Note 1).

1. Grow cultures under the optimal growth conditions for the species and on suitable media (*see* Section 2.2., items 1, 2, and 3; Note 4).
2. Prepare a spore suspension in sterile 10% (w/v) skimmed milk; 5% (w/v) inositol mixture.
3. Dispense 0.2-mL aliquots of the spore suspension into the sterilized and labeled ampules ensuring the suspension does not run down the inside of the ampule (*see* Note 15). Then cover the ampules with lint caps.
4. Transfer the ampules to the spin freeze-dryer, and spin while the chamber is evacuated. Cool the suspensions at approx 10°C/min. (This is uncontrolled, the rate depends on the amount of water and the pressure in the system.)
5. After 30 min switch off the centrifuge; the spore suspension will have frozen into a wedge tapering from the base of the ampule. This gives a greater surface area for evaporation of the liquid.
6. Dry for 3.5 h at a pressure of between 5×10^{-2} and 8×10^{-2} mbar, then raise the chamber pressure to atmospheric pressure and remove the ampules.
7. Plug the ampules with a small amount of sterilized cotton wool in a laminar airflow cabinet or a suitable microbiological safety cabinet (*see* Note 16).
8. Compress the plugs (aseptically) to 10 mm in depth with a glass or metal rod and push down to just above the tip of the slope of the freeze-dried suspension.
9. Constrict the plugged ampules using an air/gas torch just above the cotton wool plug. (The object is to ensure the glass is not drawn too thinly at the constriction and there is a sufficiently large bore left for the evacuation of air and the passage of water vapor.) Hold the ampule at each end and rotate in a narrow hot flame of the air/gas torch so it is heated evenly around an area 10 mm above the cotton wool plug; ensure that the ampule is turned back and forth through 360°. When the glass begins to become pliable, allow the flame to blow the glass in toward the center of the ampule while rotating the ampule slowly. Then stretch the ampule to give no more than a 10-mm increase in total length. This is performed by moving the open end section back and forward no more than 5 mm with equal and opposite movement of the closed end of the ampule. When the outer diameter of the constriction is approx one-third of its original diameter, the constriction should be complete. Place the ampule down onto a heat resistant mat while it is still slightly pliable; roll it on the mat so the ampule returns to a straight alignment.
10. Attach the constricted ampules to the secondary dryer, and evacuate. Incubate the secondary drying stage for about 17 h, this leaves a residual moisture of 1–2% by dry wt (*see* Note 17). The evolved water is absorbed by phosphorus pentoxide placed in the chamber below the manifolds of the secondary dryer (*see* Note 7).

11. Seal the ampules across the constriction, while still attached to the manifold and under vacuum (*see* Note 18), using an air/gas cross fire burner. Support the ampule. Project the two flames onto opposite sides of the constriction. The flames seal and subsequently cut through the glass (*see* Note 19).

    The ampule must not be allowed to pull away from the molten seal until it has separated or a long thin extension to the ampule will be made. Use the flame of the torch to melt the glass top of the ampule so it flows to form a thickened seal.
12. Test the sealed tubes with a high voltage spark tester to ensure the seal is intact. A purple to blue illumination will appear inside the ampule indicating the pressure is low enough and the seal is intact.
13. Store the ampules under appropriate conditions (*see* Note 20).
14. After 2 d storage, open sample ampules for viability and purity tests in a class II microbiological safety cabinet. Score the ampule midway down the cotton wool plug with a serrated edged glass cutting blade. Then heat a glass rod in a bunsen until red hot, and press it down onto the score; the heat should crack the tube around the mark.
15. Reconstitute the dried suspension and revive by adding 3–4 drops of sterile distilled water aseptically with a Pasteur pipet. Allow 15–20 min for absorption of the water by the spores (*see* Note 21).
16. Streak the contents of the ampule onto suitable agar medium and incubate at an appropriate growth temperature.
17. It is advisable to rehydrate and check viabilities at regular intervals.

3.3. Shelf Freeze-Drying

All culture work should be carried out in the appropriate microbiological safety cabinet (*see* Note 1).

1. Grow cultures under the optimal growth conditions for the species and on suitable media (*see* Section 2.2., items 1, 2, and 3; Note 4).
2. Prepare a spore suspension in sterile 10% (w/v) skimmed milk and 5% (w/v) inositol mixture.
3. Dispense 0.5-mL quantities into sterile 2-mL ampules.
4. Aseptically insert sterile, grooved, butyl rubber bungs into the necks of the ampules to the premolded rim so that the groove opening is above the vial lip.
5. Place the ampules on the precooled shelf (–35°C) of the freeze-dryer (*see* Note 22).
6. Place the sample temperature probe into an ampule containing the skimmed milk and inositol mixture only. When the temperature of this reaches

−20°C, evacuate the chamber; this reduces the temperature of the sample to −145°C as the latent heat of evaporation is removed and rises again to the shelf temperature.
7. Maintain the shelf temperature at −35°C for 3 h and then raise to +10°C at 0.08°C/min (*see* Note 23).
8. After drying for 24 h from the time the temperature of the sample reaches − 45°C, lower the shelf base to push the bungs into the neck of the ampules to seal them.
9. Raise the chamber pressure to atmospheric pressure and heat-seal the ampules above the preconstriction using an air/gas torch (in a similar manner to the glass ampules for cryopreservation; *see* Section 3.1., step 4) ready for storage. Retain a final vacuum of approx 4×10^{-2} mbar.
10. Test the sealed tubes with a high voltage spark tester. A purple to blue illumination will appear inside the ampule indicating the pressure is low enough and the seal is intact.
11. Store the ampules in appropriate conditions (*see* Note 20).
12. After 4 d storage, open sample ampules for viability and purity tests. Ampules are snapped open at the preconstriction in a class II microbiological safety cabinet. Add 0.3 mL of sterile distilled water and then temporarily plug the ampule with sterile cotton wool.
13. After 30 min, remove the cotton wool plug and transfer the contents of the ampule onto suitable agar medium using a Pasteur pipet. Incubate at an appropriate growth temperature.
14. It is advisable to rehydrate and check viabilities at regular intervals.

4. Notes

1. All exposure to microorganisms must be reduced to a minimum. This entails the containment of many procedures, particularly those that may create aerosols. Although good aseptic technique will contain organisms during simple transfers, it is essential that where more intricate procedures are carried out, a suitable microbiological safety cabinet is used. The latter becomes essential when hazard group 2 organisms are being handled. Hazard groups and the containment level necessary for handling them are defined in the Advisory Committee for Dangerous Pathogens, Categorisation of Pathogens According to Hazard and Categories of Containment (1990) *(24)*. The Control of Substances Hazardous to Health (COSHH) regulations (1988) and their subsequent amendments enforce these requirements in the UK *(25)*. Fungi may also produce volatile toxins which may be harmful to humans. These too must be contained or disposed of and a suitable cabinet should be used. A class II microbiological safety cabinet is recommended, as this not only protects the worker but also protects the

cultures from contamination. Fungi of the higher hazard group 3 or 4 may require total containment in a class III microbiological cabinet or glove box.
2. Sterile cultures survive the technique well but it is often best to allow full development before preservation. Sporulating cultures give better recovery. Poor isolates will not be improved by this method and may be more sensitive to the process, giving rise to preservation failures.
3. Glass ampules are not always easy to handle; they may crack and it is difficult to avoid flaws in the seals. Faulty ampules allow liquid nitrogen to enter and may explode on removal from the refrigerator when the liquid vaporizes. However, at IMI only a small number of glass ampules have "exploded" in 20 yr of use. It is imperative that the seals are sound and thick. The ampules should be placed in a metal container when removed from storage and held in the vapor phase for several seconds to allow any liquid nitrogen to evaporate. This problem can be avoided by storage in the vapor phase. Various plastic ampules have been developed; these are usually stored in the vapor phase to avoid seepage of nitrogen through the cap seal.
4. Only sporulating fungi seem to survive centrifugal freeze-drying, though some sterile ascocarps, sclerotia, and other resting stages have been processed successfully. However, the method of shelf freeze-drying can be much more successful for these than the spin freeze method. Cooling rate, drying temperature, and rate of heat input can be optimized for the organism. Organisms, which have failed spin freeze-drying, have survived this method *(22)*.
5. The skimmed milk and inositol mixture is sensitive to heat and denatures easily. The sugars are caramelized, therefore the temperature and the time of exposure must be controlled. Autoclave at 121°C for 10 min.
6. Lint caps, fluffy side innermost can be made to go over individual tubes or batches of tubes. At IMI, 15 replicate ampules are covered by one cap. The tubes are placed in metal racks covered with aluminum foil and sterilized.
7. The phosphorus pentoxide desiccant is harmful and caustic; all contact must be avoided. Desiccant trays should be filled carefully to avoid bringing the powder into the air. Goggles, gloves, and particle masks should be worn.
8. Sterilization of the butyl rubber bungs by autoclaving at 121°C for 15 min will introduce water into the bung that will be liberated during the freeze-drying process or storage afterward. Immersion in sterile industrial methylated (IMS) will help remove the water and can be evaporated away. Alternatively, surface sterilization with IMS without autoclaving can be sufficient to prevent contamination of the freeze-dried product.

9. Cold hardening of the cultures prior to freezing may prove beneficial. Pregrowth of cultures in the refrigerator (4–7°C) can improve postthaw viabilities of some fungi, though others cannot grow at low temperatures. For those isolates that are sensitive, a short exposure to these temperatures may be beneficial, or this stage can be omitted altogether.
10. Various precautions can be taken to prevent mechanical damage. Fungi on slivers or blocks of agar can be placed in the ampules, or the fungus can be grown on small amounts of agar in the ampule before the cryoprotectant is added (plastic cryotubes are more suitable for this). An alternative is to grow cultures on plant seeds in liquid culture or on small inanimate particles. These can then be transferred to the cryotube and frozen.
11. Cryoprotectants protect in several different ways but they must be allowed to come to equilibrium with the cells. The permeable protectants must be given time to enter the cell and this time depends on the permeability of the cell membrane. Generally the cell membrane of fungi is more permeable to DMSO than to glycerol. Larger molecular weight substances, such as sugars and PVP, do not penetrate the cell and either protect by reducing the amount of water in the cell through exosmosis or impede ice crystal formation. Normally a period of at least 1 h is necessary for equilibration.
12. Liquid nitrogen is considered to be a hazardous substance. Since it is extremely cold it will produce injuries similar to burns and it is also an asphyxiant gas at room temperature. It is important that it is handled with care. Contact must be avoided with the liquid or anything it has come into contact with. A face shield should be worn to prevent splashes hitting the face. Cold resistant gloves will prevent direct contact, but care must be taken not to allow nitrogen to splash into them. The liquid can penetrate the ampules and vials during storage. On retrieval of the ampule from liquid nitrogen the liquid will expand, and if it can not escape, the ampule will explode. Plastic ampules may split or the cap may fly off, whereas glass ampules may shatter and therefore be more hazardous. The liquid and culture storage vessels must be stored in a well ventilated area and it is recommended that the level of oxygen in the area be monitored. If the level of oxygen in the atmosphere falls below 18% (v/v), anyone present will suffer drowsiness and headache. If the level falls to 16% (v/v), this is potentially lethal. As the generated nitrogen is initially cold there will be a higher concentration closer to the ground. If a person faints he or she would be in danger of asphyxiation. At IMI the liquid nitrogen storage area has a low level oxygen alarm and, in other areas where liquid nitrogen is used, staff members are issued with an individual oxygen monitor.

13. A reproducible method of cooling to the holding temperature can be achieved in a controlled rate cooler. IMI uses a Kryo 16 programmable cooler (Planer Products Ltd., Sunbury on Thames, Middlesex, UK). The control temperature is measured in the chamber wall and therefore the sample temperature can vary quite widely from the programmed protocol. The program must take this into account and be adjusted until linear cooling of the sample is achieved.
14. A controlled rate cooler can be programmed to thaw the frozen fungi. The ampules are heated at about 200°C/min, raising the chamber temperature to +50°C. The sample is removed before its temperature reaches 20°C.
15. When filling the ampules it is important not to allow the suspension to run down the length of its inner surface. When the ampule is heated during constriction the suspension burns releasing fumes that may be toxic to the freeze-dried material and leave residues that will interfere with the eventual sealing of the ampule.
16. The period when ampules are kept at atmospheric pressure between drying stages must be as short as possible as the exposure to the atmosphere of the partly dried material can cause deterioration *(26)*.
17. Overdrying will kill or at least cause mutation by damaging DNA *(27)*, whereas having too high a residual moisture will allow rapid deterioration during storage *(22)*. A residual water content between 1 and 2% by dry weight proves successful for fungi.
18. At IMI, evacuation continues while sealing after the second stage drying to ensure low pressure levels in the ampule and therefore good storage conditions. An alternative method is to back fill with a dry inert gas such as nitrogen or argon.
19. If the flame is allowed to heat the ampule to either side of the constriction during sealing, the molten glass will be pushed in by atmospheric pressure and the ampule may implode.
20. Storing the ampules at a low temperature is thought to give greater longevities and 4°C seems to be favored *(18)*. At IMI the ampules are stored at temperatures between 15 and 20°C, and fungi have survived over 25 yr *(1,6)*.
21. Rehydration of the fungi should be carried out slowly giving time for the absorption of moisture before plating on to a suitable medium. It is sometimes necessary to rehydrate in a controlled environment for some very sensitive strains *(28)*.
22. The shelf temperature of a shelf freeze-drier can be controlled to cool at a particular rate. This enables the cooling stage of the freeze-drying process to be optimized for individual fungal strains. However, it may be prefer-

able to precool the ampules in a programmable cooler and then transfer them onto precooled shelves.
23. The temperature of the sample must be kept below its melting point during drying. The shelf temperature is therefore kept low (–35°C) during the initial stages of the process. The freezing point of fungal cytoplasm is quite often between –15 and –20°C *(22)*, and therefore the temperature must remain below this until all unbound water is removed. The suspension reaches 5% moisture content after 3 h, the warming protocol takes over 6 h to rise to –20°C.

References

1. Smith, D. and Onions A. H. S. (1983) *The Preservation and Maintenance of Living Fungi.* Wallingford, UK, CAB Mycological Institute.
2. Calcott, P. H. (1978) *Freezing and Thawing Microbes.* Meadowfield Press, Shildon, UK.
3. Franks, F. (1981) Biophysics and biochemistry of low temperatures and freezing, in *Effects of Low Temperature of Biological Membranes* (Morris, G. J. and Clarke A., eds.), Academic, London, pp. 3–19.
4. Smith, D. (1993) Tolerance to freezing and thawing, in *Stress Tolerance of Fungi* (Jennings, D. H., ed.), Marcel Dekker, New York, pp. 145–171.
5. Morris, G. J. (1981) *Cryopreservation: An Introduction to Cryopreservation in Culture Collections.* Culture Centre of Algae and Protozoa, ITE, Cambridge, UK.
6. Kirsop, B. E. and Doyle, A. (1991) *Maintenance of Microorganisms and Cultured Cells: A Manual of Laboratory Methods.* Academic, London.
7. Smith, D. (1988) Culture and maintenance, in *Living Resources for Biotechnology: Filamentous Fungi* (Hawksworth, D. L. and Kirsop, B. E., eds.), Cambridge University, Cambridge, UK, pp. 75–99.
8. Hwang, S.-W. (1960) Effects of ultralow temperature on the viability of selected fungus strains. *Mycologia* **52**, 527–529.
9. Polge, C., Smith, A. U., and Parkes, S. (1949) Revival of spermatozoa after dehydration at low temperatures. *Nature (Lond.)* **164**, 666.
10. Morris, G. J., Smith, D., and Coulson, G. E. (1988) A comparative study of the morphology of hyphae during freezing with the viability upon thawing of 20 species of fungi. *J. Gen. Microbiol.* **134**, 2897–2906.
11. Onions, A. H. S. (1983) Preservation of fungi, in *The Filamentous Fungi 4, Fungal Technology* (Smith, J. E., Berry, D. R., and Kristiansen, B., eds.), Edward Arnold, London, pp. 375–390.
12. Hwang, S.-W. and Howell, A. (1968) Investigation of ultra-low temperature of fungal cultures II. Cryoprotection afforded by glycerol and dimethyl sulfoxide to eight selected fungal cultures. *Mycologia* **60**, 622–626.
13. Hwang, S.-W., Kylix, W. F., and Haynes, W. C. (1976) Investigation of ultra low temperature for fungal cultures III. Viability and growth rate of mycelial cultures cryogenic storage. *Mycologia* **68**, 377–387.
14. Ashwood-Smith, M. J. and Warby, C. (1971) Studies on the molecular weight and cryoprotective properties of PVP and Dextran. *Cryobiology* **8**, 453–464.

15. Smith, D. (1983) Cryoprotectants and the cryopreservation of fungi. *Trans. Br. Mycol. Soc.* **80**, 360–363.
16. Hwang, S.-W. (1966) Long term preservation of fungus cultures with liquid nitrogen refrigeration. *Appl. Microbiol.* **14**, 784–788.
17. Hwang, S.-W. (1968) Investigation of ultra low temperature for fungal cultures I. An evaluation of liquid nitrogen storage for preservation of selected fungal cultures. *Mycologia* **60**, 613–621.
18. Heckly, R. J. (1978) Preservation of microorganisms. *Adv. Appl. Microbiol.* **24**, 1–53.
19. Raper, K. B. and Alexander, D. F. (1945) Preservation of moulds by the lyophil process. *Mycologia* **37**, 499–525.
20. Jong, S. C., Levy, A., and Stevenson, R. E. (1984) Life expectancy of freeze-dried fungus cultures stored at 4°C, in *Proceedings of the Fourth International Conference on Culture Collections* (Kocur, M. and daSilva, E., eds.), World Federation of Culture Collections, London, pp. 125–136.
21. von Arx J. A. and Schipper, M. A. A. (1978) The CBS fungus collection. *Adv. Appl. Microbiol.* **24**, 215–236.
22. Smith, D. (1986) *The Evaluation and Development of Techniques for the Preservation of Living Fungi.* Thesis submitted to London University for the degree of Doctor of Philosophy.
23. Smith, D. and Kolkowski, J. (1992) Fungi, in *Preservation and Maintenance of Cultures Used in Biotechnology and Industry,* Butterworth, Stoneham, MA.
24. Advisory Committee for Dangerous Pathogens (ed.) (1990) *Categorisation of Pathogens According to Hazard and Categories of Containment.* Health and Safety Executive, HMSO, London.
25. The Control of Substances Hazardous to Health (ed.) (1988) *Regulations.* Health and Safety Commission, London.
26. Rey, L. R. (1977) Glimpses into the fundamental aspects of freeze drying, in *Development in Biological Standardisation.* International symposium of freeze drying of biological products (Cabasso, V. J., Regamey, R. H., and Krager S., eds.), Basel, Switzerland, pp. 19–27.
27. Ashwood-Smith, M. J. and Grant, E. (1976) Mutation induction in bacteria by freezedrying. *Cryobiology* **13**, 206–213.
28. Staffeld, E. E. and Sharp, E. L. (1954) Modified lyophil method for preservation of *Pythium* species. *Phytopathology* **44**, 213,214.

CHAPTER 7

Cryopreservation of Pathogenic and Nonpathogenic Free-Living Amoebae

Simon Kilvington

1. Introduction

Free-living amoebae (FLA) are unicellular, eukaryotic protozoa found in virtually all soil and aquatic environments *(1)*. All FLA are characterized by a feeding and replicating trophozoite form that, in most genera, can produce a resistant cyst stage in response to adverse conditions *(1)*. FLA claim our interest not only as fundamentally interesting microbes about which there is still much to be learned, but also because of the capacity for certain species to cause serious and often fatal infections in humans. The amoeboflagellate *Naegleria fowleri* is found in thermally polluted water and causes primary amoebic meningoencephalitis in previously healthy persons *(2–4)*. Members of the genus *Acanthamoeba* are among the most common FLA and some species cause granulomatous amoebic encephalitis in the immunocompromised host and a destructive eye infection in previously healthy persons termed *Acanthamoeba* keratitis *(2,4)* (*see* Note 1).

Research at the Bath Public Health Laboratory has centered on the development of improved methods for the isolation, identification, and taxonomic classification of members of the genera *Naegleria* and *Acanthamoeba*. These studies have required the accumulation of many hundreds of strains from human and environmental sources worldwide. Although the trophozoites of these amoebae can be easily grown on agar plates spread with a bacterial lawn (monoxenic culture) and also be adapted to axenic (bacteria-free) growth in semidefined broth media, it is impractical to maintain large numbers of strains by continuous culture. Beside the expensive and labor-intensive nature of the process, cultures can also be mislabeled or contaminated by regular subculturing. Serial subculture, particularly of axenic strains, can also result in the loss of biological characteristics such as virulence and ability to encyst or flagellate *(3,5,6)*.

This prompted the investigation of methods for the long-term cryopreservation of both pathogenic and nonpathogenic FLA. Based on a method described previously for the cryopreservation of the anaerobic parasite *Trichomonas vaginalis (7)*, it was found that the trophozoites of *Naegleria*, *Acanthamoeba*, and other FLA could be successfully cryopreserved from axenic or monoxenic cultures using a two-step cooling process in laboratory freezers *(8)*. Dimethyl sulfoxide (DMSO) is used as the cryoprotectant and cooling achieved by placing the ampules directly at –20°C for 60 min, followed by another 60 min at –70°C. The ampules are then either held at –70°C or transferred to liquid nitrogen. Although relatively low recovery rates are obtained, indicating that conditions are not optimal, the method is proposed as a simple and reproducible means of cryopreserving pathogenic and nonpathogenic FLA.

2. Materials

1. Nonnutrient agar-*Escherichia coli* medium (NNA-*E. coli*): 1.5% (w/v) agar in 25% strength Ringer's solution. Autoclave at 121°C for 15 min, cool to 50°C and pour plates (22–25/500 mL). Dry plates overnight at 37°C and store at 4°C for up to 2 wk.

    Spread a single colony of *E. coli* (any strain) on to a nutrient agar plate and incubate overnight at 37°C. Store plate at 4°C and subculture weekly. Recover all the bacteria with a cotton-tipped swab and suspend in 2 mL of 25% strength Ringer's or distilled water. Using a Pasteur pipet, inoculate three drops of the *E. coli* suspension on to the center of an NNA plate. Spread the *E. coli* evenly over the surface of the agar using a bent glass rod or cotton-tipped swab. Store NNA-*E. coli* plates at 4°C for use within 1 wk (*see* Note 2).

2. Modified Chang's serum-casein-glucose-yeast extract medium (SCGYM):
    a. Dissolve the following in distilled water and make up to 890 mL: 1.325 g Na_2HPO_4, 0.8 g KH_2PO_4, 10 g casein digest, 5 g yeast extract, 2.5 g glucose. Adjust to pH 7.0 if necessary. Autoclave at 121°C for 15 min and store at 4°C.
    b. Dissolve 10 g of Panmede liver digest in 90 mL of distilled water. Adjust to pH 7.0, make up to 100 mL, and filter sterilize. Distribute in 10–mL vol and store frozen at –20°C.

c. For use add 10 mL of b to 890 mL of a, 100 mL of fetal calf serum (heat inactivated), and 2.5 mL each of 10,000 U/mL penicillin and streptomycin. Store complete medium at 4°C and use within 2 wk (*see* Note 2).
3. DMSO cryopreservation solution: To 9 mL of SCGYM base (Section 2., items 2a and 2b), add 1 mL of DMSO (10% [v/v] final concentration) in a polypropylene tube (DMSO dissolves polystyrene). Mix thoroughly and store at 4°C in the dark for use within 1 wk (*see* Note 2).
4. 25% Strength Ringer's solution: Dissolve one tablet of Ringer's salts in 500 mL of distilled water, dispense in 100-mL vol and autoclave at 121°C for 15 min. Store at room temperature.
5. Two freezers set at −20 and −70°C.
6. Liquid nitrogen storage container and aluminum canes.
7. Safety equipment: Cryogloves, goggles, and so on.
8. Laminar flow cabinet and appropriate containment facilities.

3. Methods

As some FLA are pathogens of humans and other animals, appropriate precautions should be observed (*see* Note 1).

3.1. Axenic Trophozoites

1. Chill a tissue-culture tube containing a semiconfluent growth of trophozoites on ice for 10 min and harvest the amoebae by centrifugation at 500g for 5 min at room temperature (*see* Note 3).
2. Resuspend the cell pellet in 2 mL of complete SCGYM (*see* Section 2., item 2) at room temperature and aliquot in 0.5-mL vol into 1.2 mL, screw-capped, polypropylene ampules measuring 41 × 12 mm.
3. Add 0.5 mL of the DMSO cryopreservation solution (*see* Section 2., item 3), equilibrated to room temperature, to each of the ampules and mix by inversion three times (*see* Note 4).
4. Immediately place the ampules upright into the bottom of a −20°C freezer and incubate for 60 min (*see* Note 5).
5. Transfer the ampules to a −70°C freezer and leave for a further 60 min (*see* Note 5).
6. Either store the ampules at −70°C (*see* Note 6) or attach them to aluminum canes and plunge into liquid nitrogen.
7. To recover the cryopreserved trophozoites, remove an ampule from the liquid nitrogen or −70°C freezer and place into an enclosed container under an operating microbiological safety cabinet and leave for 3 min (*see* Note 1).
8. Place the ampule in a 37°C water bath and invert occasionally to hasten thawing.
9. When thawed, inoculate the ampule contents into 3 mL of prewarmed complete SCGYM and incubate for 60 min.

10. Gently decant the liquid from the culture tubes and replace with 3 mL of prewarmed complete SCGYM and incubate under appropriate culture conditions (*see* Notes 7–10).

3.2. Monoxenic Trophozoites

1. Flood the NNA-*E. coli* plate with 5 mL of 25% strength Ringer's solution and gently suspend the trophozoites by rubbing the agar surface with a bent glass rod or cotton-tipped swab (*see* Note 3).
2. Pipet the trophozoites into a screw-capped tube and centrifuge at 500g for 5 min at room temperature.
3. Resuspend the cell pellet in 1 mL of complete SCGYM at room temperature and aliquot in 0.5-mL vol into 1.2 mL, screw-capped, polypropylene ampules.
4. Follow steps 3–6 of Section 3.1.
5. To recover the trophozoites follow steps 7 and 8 of Section 3.1.
6. When thawed, inoculated approx 0.5 mL on to the surface of a NNA-*E. coli* plate. Allow the liquid to absorb and incubate (*see* Notes 7–10).

3.3. Cysts

1. Flood the NNA-*E. coli* plate with 5 mL of 25% strength Ringer's solution and suspend cysts by rubbing the agar surface with a bent glass rod or cotton-tipped swab (some effort may be required to dislodge the cysts) (*see* Note 3).
2. Pipet the cysts into a screw-capped tube and centrifuge at 500g for 5 min at room temperature.
3. Resuspend the cell pellet in 1 mL of 25% strength Ringer's solution and aliquot in 0.5-mL vol into 1.2-mL, screw-capped, polypropylene ampules.
4. Add 0.5 mL of the DMSO cryopreservation solution, equilibrated to room temperature, to each of the ampules and mix by inversion three times.
5. Place upright into a –20°C freezer for storage at this temperature.
6. To recover the cysts, thaw an ampule in a 37°C circulating water and spread approx 0.5 mL on to the surface of a NNA-*E. coli* plate. Allow the liquid to absorb and incubate (*see* Notes 1, 7, and 11).

4. Notes

1. *N. fowleri* and many *Acanthamoeba* are pathogenic to humans. Therefore, all manipulations of these organisms should be performed under appropriate containment facilities using a microbiological safety cabinet (*9*). Liquid nitrogen can also penetrate the washer sealing between the ampule cap and tube and produce an explosion on warming. Ampules should therefore be placed inside an enclosed container for 3 min immediately after removal from liquid nitrogen.
2. All chemicals used for the culture and cryopreservation of FLA are analytical grade. Growth media are from Difco Laboratories (Surrey, UK) and

the Panmede liver digest from Paines & Byrne Ltd. (Greenford, UK). Oxoid (Basingstoke, UK) supply tablets for the preparation of 25% strength Ringer's solution. Heat-inactivated fetal calf serum, Petri dishes, flat-sided tissue-culture tubes, and cryopreservation ampules are available from Gibco (Middlesex, UK). All procedures are performed under aseptic conditions using sterile reagents and plasticware.

3. A tissue-culture tube containing an almost confluent growth of amoebae in SCGYM contains approximately 5×10^5 to 1×10^6 trophozoites/mL and is sufficient for preparing four ampules for cryopreservation. Yields of trophozoites and cysts from NNA-*E. coli* medium are much lower and only two ampules should be prepared per culture plate.

4. Although DMSO at a concentration of 5–10% (v/v) is the most widely used protectant for the cryopreservation of both free-living and parasitic protozoa, other agents such as glycerol and methanol, have been used *(10,11)*. In addition, an equilibration period for the cells in the presence of the cryoprotectant at different temperatures is often employed *(10)*. Studies addressing these variables are therefore indicated and may also lead to improved viable recovery rates.

5. Freezers used for cryopreservation should not be opened for at least 1 h before use and should remain closed during the cooling steps. Ampules should be transported between the freezers and liquid nitrogen tank with the minimum delay and inside a small enclosed polystyrene box to prevent warming.

6. As well as *Naegleria* and *Acanthamoeba*, FLA of the genera *Hartmannella*, *Vahlkampfia*, and *Willaertia* have also been successfully cryopreserved by the techniques described here with viable recovery after 4 yr storage in liquid nitrogen and 2 yr at –70°C. This suggests the methods may be suitable for the cryopreservation of many other genera of FLA. The methods are simple and reproducible and require equipment that may be considered standard in most laboratories. Importantly, the problems associated with the maintenance of large numbers of strain by continuous culture are eliminated.

7. *Naegleria*, *Acanthamoeba*, and most other FLA grow well on NNA-*E. coli* medium, producing zones of dense trophozoite growth radiating away from a central inoculum on a culture plate. Cysts are formed during prolonged incubation (5–7 d) of the plates. NNA-*E. coli* culture plates are incubated in sealed polythene bags to prevent desiccation. The temperature of incubation depends on the species of FLA. All pathogenic species will grow at 37°C or above and most others at 32°C *(1)*. *Naegleria* and most *Acanthamoeba* can be adapted to axenic culture from NNA-*E. coli* plates, although several attempts may be necessary. An area of agar (1×2.5 cm)

containing dense trophozoite growth is cut from the plate and inoculated into a tube of SCGYM. Lower temperatures are usually required for successful axenization. Culture tubes should therefore be incubated at 32°C for strains tolerating growth at 37°C on NNA-*E. coli* and room temperature for all others. Change the medium and remove the agar block after 48 h incubation.

8. A study of the recovery rates of *Naegleria* and *Acanthamoeba* trophozoites cryopreserved in liquid nitrogen by the methods detailed here gave: *N. fowleri* 8% ± 3%, *N. lovaniensis* 12% ± 3%, *N. gruberi* 25% ± 6%, and *A. polyphaga* 34% ± 5.5% from axenic cultures, and 21% ± 7%, 31% ± 10%, 44% ± 11%, and 49% ± 9%, respectively, from monoxenic cultures *(8)*. Although these figures are low, particularly for axenic cultures of *N. fowleri* and *N. lovaniensis*, confluent growth of trophozoites are usually obtained within 2–3 d incubation.

9. The relatively low recovery rates following cryopreservation indicate that the methods are not optimal for FLA. The rate of cooling is an important factor in the cryopreservation of free-living and parasitic protozoa, with values of approx 1°C/min to –60 or –70°C seeming to be optimal *(10–13)*. This can be achieved using programmable freezers *(10,13)* or by suspending ampules in a holding device in the vapor phase of liquid nitrogen in a dewar vessel *(12,14)*. Improved results may therefore be obtained if the cooling rate could be controlled at 1°C/min to –70°C using such methods.

10. Low recovery rates do result in a selection pressure on the original cell population used for cryopreservation. However, this is not necessarily a problem with FLA as there is no evidence for genetic exchange or sexual reproduction in these organisms and clonal cell populations for cryopreservation can be easily derived from single cysts or trophozoites *(1)*. Furthermore, we have been unable to detect changes in isoenzyme profiles, DNA restriction fragment length polymorphisms, or ability to transform into the flagellate stage in strains of *Naegleria* spp. that have been repeatedly cryopreserved and recovered over several years.

11. The method for the cryopreservation of FLA cysts at –20°C is based on a previously described method *(15)* for *Acanthamoeba* except that 5% DMSO is used instead of 1.5*M* methyl alcohol. Although not extensively used in this laboratory, *Naegleria* and *Acanthamoeba* strains cryopreserved by this method have been successfully recovered after 1–2 yr storage.

References

1. Page, F. C. (1988) *A New Key to Freshwater and Soil Gymnamoebae.* Freshwater Biological Association, Ambleside, Cumbria, UK.
2. Warhurst, D. C. (1985) Pathogenic free-living amoebae. *Parasitol. Today* **1**, 24–28.

3. Marciano-Cabral, F. (1988) Biology of *Naegleria* spp. *Microbiol. Rev.* **52**, 114–133.
4. Ma, P., Visvesvara, G. S., Martinez, A. J., Theodore, F. H., Dagget, P. M., and Sawyer, T. K. (1990) *Naegleria* and *Acanthamoeba* infections: review. *Rev. Infect. Dis.* **12**, 490–513.
5. Stevens, A. R. and O'Dell, W. (1974) *In vitro* growth and virulence of *Acanthamoeba. J. Parasitol.* **60**, 884,885.
6. Wong, M. M., Karr, S. L., and Chow, C. K. (1977) Changes in the virulence of *Naegleria fowleri* maintained *in vitro. J. Parasitol.* **63**, 872–878.
7. Miyata, A. (1975) On the cryo-biological study of the parasitic protozoa. 3. Effects of temperature and time of equilibration with glycerol or DMSO on survival of *Trichomonas vaginalis. Trop. Med. (Nagasaki)* **17**, 55–64.
8. Kilvington, S. and White, D. G. (1991) A simple method for the cryopreservation of free-living amoebae belonging to the genera *Naegleria* and *Acanthamoeba. Europ. J. Protistol.* **27**, 115–118.
9. Advisory Committee on Dangerous Pathogens (1990) *Categorisation of Pathogens According to Hazard and Categories of Containment.* HMSO, London.
10. James, E. R. (1984) Maintenance of parasitic protozoa by cryopreservation, in *Maintenance of Microorganisms: A Manual of Laboratory Methods* (Kirsop, B. E. and Snell, J., eds.), Academic, London, pp. 161–176.
11. Leeson, E. A., Cann, J. P., and Morris, G. J. (1984) Maintenance of algae and protozoa, in *Maintenance of Microorganisms: A Manual of Laboratory Methods* (Kirsop, B. E. and Snell, J., eds.), Academic, London, pp. 131–160.
12. Warhurst, D. C. and Wright, S. G. (1979) Cryopreservation of *Giardia intestinalis. Trans. R. Soc. Trop. Med. Hyg.* **72**, 601.
13. Farri, T. A., Warhurst, D. C., and Marshall, T. F. de C. (1983) The use of infectivity titrations for measurement of the viability of *Entamoeba histolytica* after cryopreservation. *Trans. R. Soc. Trop. Med. Hyg.* **77**, 259–266.
14. Kasten, F. H. and Yip, D. K. (1976) A simple device and procedure for successful freezing of cells in liquid nitrogen vapor. *Methods Cell Biol.* **14**, 165–179.
15. Badenoch, P. R., Johnson, A. M., Christy, P. E., and Coster, D. J. (1990) Pathogenicity of *Acanthamoeba* and a corynebacterium in the rat cornea. *Arch. Ophthalmol.* **108**, 107–112.

CHAPTER 8

Long-Term Cryopreservation of Thylakoid Membranes

Dirk K. Hincha and Jürgen M. Schmitt

1. Introduction

Chloroplast thylakoids are the site of photosynthetic light reactions, electron transport, adenosine triphosphate (ATP) synthesis and nicotinamide adenine dinucleotide phosphate (NADP) reduction in plant cells. Isolated thylakoids are therefore the necessary prerequisite for many in vitro studies of photosynthetic mechanisms. They are also used to elucidate aspects of photosynthetic responses to freezing, chilling, heat, and high light stress. For many of these studies it is important to work with membranes of closely similar properties over extended periods of time to facilitate direct comparisons of the results. The pigment, protein, and lipid composition of thylakoid membranes, however, vary with plant growth conditions *(1–3)*. Even the solute permeability properties of thylakoids are significantly different between warm- and cold-grown spinach plants *(4)*.

One way of ensuring homogeneous composition of the membranes for longer time periods is the frozen storage of batches of thylakoids from one preparation. This requires storage conditions that minimize the damaging effects of freezing and thawing. For thylakoids, short-term cryopreservation (several hours at –20°C) can be achieved without problems in the presence of high concentrations of sugars (e.g., 0.5M sucrose; *see [5]* for a review). At lower sugar concentrations, thylakoids rupture during thawing. This is indicated by the release of the electron transport protein plastocyanin from the vesicle lumen (*[6];* Fig. 1A). At

From: *Methods in Molecular Biology, Vol. 38: Cryopreservation and Freeze-Drying Protocols*
Edited by: J. G. Day and M. R. McLellan Copyright © 1995 Humana Press Inc., Totowa, NJ

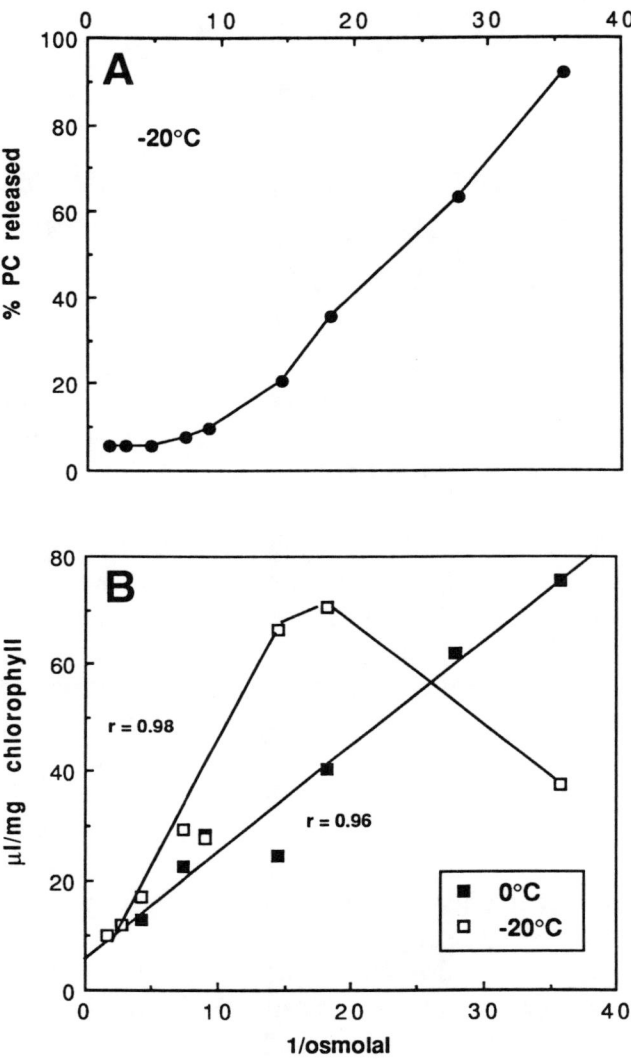

Fig. 1. Plastocyanin (PC) release (**A**) and thylakoid volume (**B**) as a function of the reciprocal osmolalities of the suspending media (Boyle-van't Hoff plot) after 3 h at 0°C or −20°C. The membranes were washed in 5 mM NaCl. The samples contained 2.5 mM NaCl and sucrose between 20 mM (1/osmolal = 35.7) and 500 mM (1/osmolal = 1.6). The straight lines in (B) were fitted to the data by linear regression analysis. For greater clarity, the calculated nonosmotic volume has been subtracted from the measured data.

the same time thylakoids lose other proteins and osmotically active solutes (*[7];* Fig. 1B). This leads to an inactivation of light-driven electron transport and ATP synthesis *(5)*. At higher sugar concentrations, plastocyanin release is prevented, but the vesicles show an increased volume when compared to unfrozen controls (Fig. 1). This increase in volume is caused by the diffusion of solutes across the membranes in the frozen state *(7)*. Solute loading of the vesicles increases with storage time, resulting in rupture when critical membrane tension is exceeded. Rupture events are a complex function of time, freezing temperature, and solute concentration *(8)*.

When different sugars were compared for their ability to prevent freeze–thaw damage to thylakoids, significant differences in their molar effectiveness were found *(9–12)*. This could be related to the ability of the sugars to reduce solute loading of the vesicles during freezing *(13,14)*. One of the most effective sugars in this regard was the disaccharide trehalose *(13)*. It has been shown recently that trehalose reduces the solute permeability of thylakoids, presumably by hydrogen bonding to galactolipid headgroups in the membranes *(15)*. Therefore, the addition of trehalose to a cryopreservation solution for thylakoids is expected to greatly reduce solute loading and the resulting vesicle rupture after long-term storage. In the presence of sucrose (200 mM) and trehalose (200 mM), no detectable loss of plastocyanin was observed from thylakoids isolated from nonhardy spinach plants after storage at –20°C for 3 mo.

Although sugars are cryoprotective, the storage of thylakoids in solutions not containing any inorganic ions leads to the loss of proteins from the membranes *(16)* and reduced biochemical activities. On the other hand, the addition of high concentrations of inorganic salts to a cryopreservation solution can severely damage thylakoids during a freeze–thaw cycle, or even at 0°C *(5)*. Cryopreservation is strongly influenced by the ratio of cryotoxic to cryoprotective solutes. The presence of NaCl will lead to injury when a molar ratio of salt to sugar of about 1:5 is exceeded. A further increase in salt concentration will lead to severe injury, with a total breakdown of some biochemical activities at ratios of about 5:1 *(17)*. This type of freeze–thaw damage will not result in vesicle rupture and release of plastocyanin. Instead, peripheral membrane proteins are solubilized during freezing *(6)*. This can be detected as a reduction in cyclic photophosphorylation activity, as the CF_1 part of the adenosine triphosphatase (ATPase) is among the released proteins *(5)*.

2. Materials

1. Plants: Spinach and peas are commonly used species for experiments with isolated thylakoids or chloroplasts. The methods described in this chapter were all developed with spinach thylakoids. Leaves from greenhouse or field grown plants or leaves purchased at local markets have all been used with comparable results (*see* Note 2).
2. Homogenization butter: 160 mM NaCl, 240 mM sucrose, 1 mM MgCl$_2$, 2 mM EDTA, 1 mM KH$_2$PO$_4$, 50 mM Tris (pH 7.8 with HCl), 1M Na-ascorbate, and 1M cysteine. (These stock solutions can be stored frozen at –20°C for extended times.)
3. Blender.
4. Miracloth or 50-µm nylon mesh.
5. Washing solution: 10 mM MgCl$_2$ (*see* Notes 3 and 4).
6. Chlorophyll determination: 80% (v/v) acetone. Photometer.
7. Cryoprotectant solution: 0.4M sucrose, 0.4M trehalose.
8. Freezer: Any commercially available household freezer with a constant temperature of –20 to –25°C (*see* Note 1).
9. Cyclic photophosphorylation:
    a. Incubation solution: 50 mM KCl, 15 mM Tris-HCl, pH 8.0, 5 mM MgCl$_2$, 2.5 mM KH$_2$PO$_4$, 2 mM ADP. 3 mM Phenazine methosulfate (PMS; this substance is light sensitive and should be stored frozen at –20°C).
    b. Phosphate assay: 100 mM zinc acetate, 15 mM (NH$_4$)$_2$MoO$_4$, 1% (w/v) sodium dodecyl sulfate (SDS), 20 mM Na-acetate (pH 4.5 with HCl). The SDS should be prepared as a 10% (v/v) stock solution and should be added to the assay solution directly before use, as the SDS precipitates from this solution after a few hours.
    c. Lamps and a circulating water bath.
10. Thylakoid volume measurements: Centrifuges equipped with a rotor specially designed to hold glass capillaries are offered by several manufacturers. Hematocrit capillaries and sealing wax or plastic sealing caps are available from laboratory or hospital supply companies. Osmometer to determine the osmolality of incubation solutions to construct Boyle-van't Hoff plots (*see* Fig. 1B)
11. Determination of plastocyanin release:
    a. Antiplastocyanin antibodies. In addition to our own *(6)* there are several other published methods for the purification of plastocyanin from different plant species. Antibodies can be raised in rabbits following standard immunological protocols.
    b. Agarose: 1% (w/v) agarose (e.g., Serva Standard EEO, Heidelberg, Germany), 1% (w/v) PEG4000, 1 mM Ca-lactate, 0.1% (w/v) NaN$_3$, 1% (v/v)

Triton X-100, 25 mM Tris 190 mM glycine. (pH 8.4 is reached by mixing Tris and glycine without further adjustment.)
c. Gel-Bond film (FMC BioProducts, Rockland, ME). Puncher and template for sample wells (Pharmacia, Uppsala, Sweden). Vacuum pump.
d. Tris-buffered saline solution (TBS): 150 mM NaCl, 25 mM Tris-HCl, pH 7.5.
e. Staining solution: 50% (v/v) methanol, 10% (v/v) acetic acid, 0.2% (w/v) Coomassie brilliant blue R.
f. Destaining solution: 50% (v/v) methanol, 10% (v/v) acetic acid.

3. Method
3.1. Freezing Method

Thylakoids are sensitive to elevated temperatures, therefore, all solutions and glassware should be cooled to 0–4°C before use. Isolated membranes should always be kept on ice, preferably in dim light. All centrifugation steps take place at 4°C.

1. Thylakoid isolation: Homogenize 50 g of leaves with 100 mL of homogenization buffer in a blender for approx 10 s. Add 125 µL 1M Na-ascorbate and 340 µL 1M cysteine to the buffer immediately before use, as these substances are unstable in solution. Filter the homogenate through Miracloth or nylon mesh to remove coarse particles. Centrifuge filtrate for 5 min at 2000g. Discard supernatant and resuspend pellets in approx 50 mL of washing solution. This is most easily performed using a Pasteur pipet. Centrifuge for 5 min at 7000g and discard supernatant. Repeat this washing procedure twice. Resuspend the pellets from the last centrifugation in a minimum volume of washing solution. Mix 10 µL of the thylakoid suspension with 990 µL of 80% (v/v) acetone and centrifuge for 2 min in a benchtop centrifuge. Measure absorbance of the supernatant at 663 and 645 nm with 80% (v/v) acetone as the reference. Chlorophyll content is calculated *(18)* as:

$$(8.02 \times A_{663} + 20.2 \times A_{645}) \times 0.1 = \text{mg chlorophyll/mL}$$

Dilute the thylakoid suspension with washing buffer to a concentration of at least 1 mg chlorophyll/mL (*see* Note 5).

2. Mix 0.5 mL of the thylakoid suspension with an equal volume of cryoprotectant solution in Eppendorf tubes. Place in a freezer at –20°C. No special precautions are necessary for the freezing step. Samples are most conveniently thawed in a water bath at room temperature and should be transferred to an ice bath immediately when the ice in the tubes has melted.

3.2. Measurement of Membrane Integrity

Many biochemical and physical methods have been developed in photosynthesis research that will be sensitive to impaired membrane integrity. The following three methods have been extensively used in our laboratory to evaluate freeze–thaw damage to thylakoids.

3.2.1. Cyclic Photophosphorylation Activity

Cyclic photophosphorylation activity is measured as the fixation of inorganic phosphate into ATP *(19)*.

1. Add thylakoids equivalent to approx 50 µg of chlorophyll and PMS (final concentration, 30 µM) to the incubation solution to a final vol of 1 mL in Eppendorf tubes.
2. Illuminate with bright white light (2000 W/m^2) for 40 s. Use a circulating water bath to maintain the sample temperature below 25°C. Keep reference samples in the dark.
3. Sediment the membranes by centrifugation for 5 min at 10,000g. Dilute an aliquot of the supernatant 1:10 with distilled water and mix one part diluted sample with three parts assay solution.
4. Read the absorbance at 350 nm after 30 s *(20)*. This is critical, since ATP slowly hydrolyzes at the pH employed, leading to a drift in the absorbance reading. The reference cuvet should contain one part distilled water and three parts assay solution. From a standard curve obtained by a serial dilution of the incubation solution, the rate of fixation of inorganic phosphate into ATP can be calculated.

3.2.2. Thylakoid Volume Measurements *(7)*

1. Mix aliquots of thylakoids in cryopreservation solution with sucrose solutions ranging in concentration from 0.4–0M. Make all sucrose solutions in 5 mM MgCl$_2$ (*see* Note 5).
2. Determine the osmolality of the resulting solutions with an osmometer. These measurements are made easier if you use the cryopreservation solution diluted 1:1 with the washing solution instead of thylakoids. The final results will be the same.
3. Load hematocrit capillaries with the diluted thylakoid suspensions and seal the capillaries at one end. Centrifuge for 15 min in a hematocrit centrifuge and measure pellet heights with a magnifying glass and a 0.1-mm scale. Calculate the packed thylakoid volume and construct a Boyle-van't Hoff plot, as shown in Fig. 1B. This should yield a straight line. An increase of the slope with storage time indicates solute loading during freezing *(8)*. An extrapolation of the line to an infinite solute concentration (1/osm = 0) will

give the nonosmotic volume of the thylakoid suspension. This often varies considerably between different membrane preparations, mainly because of the varying starch content of the chloroplasts.

3.2.3. Determination of Plastocyanin Release (6)

A description of the purification of plastocyanin and the preparation of monospecific antibodies is beyond the scope of this chapter. If antibodies are available, several methods can be used for detection of plastocyanin release from thylakoids. *(See [21]* for a more detailed description of basic immunological techniques.) The most simple and convenient quantitative method for the determination of antigens is single radial immunodiffusion *(22).*

1. Remove an aliquot of the thylakoid suspension and lyse it by adding Triton X-100 to a final concentration of 2% (v/v). This total thylakoid lysate serves as a 100% plastocyanin release reference sample.
2. Then sediment the thylakoids from the remaining samples by centrifugation (15 min at 16,000g). The supernatant contains the plastocyanin that was released owing to thylakoid rupture during a freeze–thaw cycle.
3. Dissolve the agarose by boiling and let it cool in a water bath to 55°C. Use approx 200 µL of agarose solution/cm^2 of gel.
4. Mix an appropriate volume of agarose solution with the necessary amount of antiserum. The amount of antiserum necessary varies with different batches and has to be determined empirically.
5. Quickly pour the solution on a glass plate. Later handling of the gels is very much facilitated when gels are not poured directly on the glass but instead on a Gel Bond film.
6. Punch 15-µL wells into the gel with the help of a vacuum pump and the template.
7. Pipet 10-µL samples into the wells and let the antigen diffuse into the gel for 40–50 h. Store the gels in a moist container during that time.
8. Proteins not precipitated by the antibodies are removed by gently pressing the gels under several layers of filter paper and washing in TBS. This can be repeated twice for 20 min each to reduce background staining.
9. Dry the gels under a stream of warm air and then submerse in the staining solution for 10 min.
10. Destain until the precipitates are clearly visible and then again dry in warm air. Measure the radius of the circular precipitates under a magnifying glass with a 0.1-mm scale. The area enclosed by the precipitates is a direct function of the amount of antigen (in this case, plastocyanin) in the samples. By comparing it to the total thylakoid lysate, relative plastocyanin release values can be calculated.

4. Notes

1. The freezing temperature has a complex influence on freeze–thaw damage. At warmer temperatures permeation is faster and therefore the problems associated with solute loading are aggravated. At lower temperatures damage caused by other, not well-characterized, mechanisms is increased. Storage in liquid nitrogen after freezing to $-20°C$ could be a feasible way to circumvent problems of solute loading but has not been investigated systematically so far.
2. During cold hardening (either under natural conditions in the field or by exposure to 4°C for 2 wk in a growth chamber) thylakoids in leaves increase in frost hardiness. This increased hardiness is in part retained by the isolated membranes. Solute loading, for instance, occurs only at about half the rate measured in thylakoids from warm grown plants. The use of membranes from cold acclimated plants therefore should be considered.
3. Other washing solutions have been used by the authors and other researchers. They can generally be adapted to the needs of subsequent experiments. Solutions completely devoid of ions should, however, be avoided, since this compromises the physiological competence of the membranes.
4. High concentrations of inorganic ions are toxic to thylakoids. The effectiveness of the ions to inactivate the membranes and to solubilize membrane proteins follows the Hofmeister lyotropic power series. Anions can be ranked in the order of increasing cryotoxicity from fluoride → chloride → bromide → iodide.
5. Higher chlorophyll concentrations generally have a protective effect on thylakoid membranes during freezing. For the hematocrit measurements, a minimum concentration of about 0.3 mg/mL after dilution is necessary to allow for pellet sizes that give good resolution of volume changes. Therefore, higher chlorophyll concentrations allow higher dilutions and consequently a larger range of osmolalities for measurements. The presence of Mg^{2+} leads to, at least, a partial stacking of the membranes which then sediment more readily during hematocrit centrifugation. This leads to sharper boundaries of the pellets. However, even in the absence of Mg^{2+}, meaningful results can be obtained (Fig. 1B).

Acknowledgments

The authors would like to thank G. Meyer and W. Gross for critically reading the manuscript and Bundesministerium für Forschung und Technologie for financial support.

References

1. Webb, M. S. and Green, B. R. (1991) Biochemical and biophysical properties of thylakoid acyl lipids. *Biochim. Biophys. Acta* **1060,** 133–158.
2. Demmig-Adams, B. and Adams, W. W., III (1992) Photoprotection and other responses of plants to high light stress. *Annu. Rev. Plant Physiol. Plant Mol. Biol.* **43,** 599–626.
3. Anderson, J. M. (1986) Photoregulation of the composition, function, and structure of thylakoid membranes. *Annu. Rev. Plant Physiol.* **37,** 93–136.
4. Hincha, D. K. and Schmitt, J. M. (1988) Mechanical freeze–thaw damage and frost hardening in leaves and isolated thylakoids from spinach. II. Frost hardening reduces solute permeability and increases extensibility of thylakoid membranes. *Plant Cell Environ.* **11,** 47–50.
5. Hincha, D. K. and Schmitt, J. M. (1992) Freeze–thaw injury and cryoprotection of thylakoid membranes, in *Water and Life* (Somero, G. N., Osmond, C. B., and Bolis, C. L., eds.), Springer, Berlin, pp. 316–337.
6. Hincha, D. K., Heber, U., and Schmitt, J. M. (1985) Antibodies against individual thylakoid membrane proteins as molecular probes to study chemical and mechanical freezing damage in vitro. *Biochim. Biophys. Acta* **809,** 337–344.
7. Hincha, D. K. (1986) Sucrose influx and mechanical damage by osmotic stress to thylakoid membranes during an in vitro freeze–thaw cycle. *Biochim. Biophys. Acta* **861,** 152–158.
8. Bakaltcheva, I., Schmitt, J. M., and Hincha, D. K. (1992) Time and temperature dependent solute loading of isolated thylakoids during freezing. *Cryobiology* **29,** 607–615.
9. Lineberger, R. D. and Steponkus, P. L. (1980) Cryoprotection by glucose, sucrose, and raffinose to chloroplast thylakoids. *Plant Physiol.* **65,** 298–304.
10. Santarius, K. A. and Bauer, J. (1983) Cryopreservation of spinach chloroplast membranes by low-molecular-weight carbohydrates. I. Evidence for cryoprotection by a noncolligative-type mechanism. *Cryobiology* **20,** 83–89.
11. Santarius, K. A. and Giersch, C. (1983) Cryopreservation of spinach chloroplast membranes by low-molecular-weight carbohydrates. II. Discrimination between colligative and noncolligative protection. *Cryobiology* **20,** 90–99.
12. Steponkus, P. L., Garber, M. P., Myers, S. P., and Lineberger, D. R. (1977) Effects of cold acclimation and freezing on structure and function of chloroplast thylakoids. *Cryobiology* **14,** 303–321.
13. Hincha, D. K. (1989) Low concentrations of trehalose protect isolated thylakoids against mechanical freeze–thaw damage. *Biochim. Biophys. Acta* **987,** 231–234.
14. Hincha, D. K. (1990) Differential effects of galactose containing saccharides on mechanical freeze–thaw damage to isolated thylakoid membranes. *Cryo-Lett.* **11,** 437–444.
15. Bakaltcheva, I., Williams, W. P., Schmitt, J. M., and Hincha, D. K. (1994) The solute permeability of thylakoid membranes is reduced by low concentrations of trehalose as a co-solute. *Biochim. Biophys. Acta.* **1189,** 38–44.

16. Wolter, F. P., Schmitt, J. M., Bohnert, H. J., and Tsugita, A. (1984) Simultaneous isolation of three peripheral proteins—a 32 kDa protein, ferredoxin NADP+ reductase and coupling factor—from spinach thylakoids and partial characterization of a 32 kDa protein. *Plant Sci. Lett.* **34,** 323–334.
17. Mollenhauer, A., Schmitt, J. M., Coughlan, S., and Heber, U. (1983) Loss of membrane proteins from thylakoids during freezing. *Biochim. Biophys. Acta* **728,** 331–338.
18. Arnon, D. J. (1949) Copper enzymes in isolated chloroplasts. Polyphenoloxidase in *Beta vulgaris. Plant Physiol.* **24,** 1–15.
19. Hincha, D. K. and Schmitt, J. M. (1985) Mechanical and chemical injury to thylakoid membranes during freezing in vitro. *Biochim. Biophys. Acta* **812,** 173–180.
20. Bencini, A. D., Shanley, M. S., Wild, J. R., and O'Donovan, G. A. (1983) New assay for enzymatic phosphate release: application to aspartate transcarbamylase and other enzymes. *Anal. Biochem.* **132,** 259–264.
21. Manson, M. M. (ed.) (1992) *Methods in Molecular Biology,* vol. 10: *Immunochemical Protocols.* Humana, Totowa, NJ.
22. Mancini, G., Carbonara, A. O., and Heremans, J. F. (1965) Immunochemical quantitation of antigens by single radial immunodiffusion. *Immunochemistry* **2,** 235–254.

CHAPTER 9

Cryopreservation of Algae

John G. Day and Mitzi M. DeVille

1. Introduction

Eukaryotic microalgae and prokaryotic cyanobacteria (blue-green algae) have traditionally been maintained by serial subculture, with the frequency of transfer being largely determined by the growth characteristics of the strain. This technique has the dual disadvantages of being labor intensive and carrying risk of loss in genetic stability of the strain.

Compared with other groups of microorganisms, relatively little research has been carried out on the development of long-term preservation methods for microalgae and cyanobacteria. Freeze-drying has not been found to be a successful biostorage method for microalgae, with very low levels of viability (<1% of original population) *(1,2)* and a further reduction in viability on prolonged storage *(3,4)*. The successful lyophilization of the cyanobacterium *Nostoc muscorum*, using a method similar to that detailed in Chapter 3, has been reported, with no observed reduction in viability after 5 yr storage *(5)*. This technique has been adopted by a small number of researchers to preserve selected cyanobacterial strains.

Cryopreservation has been successfully employed to maintain algae *(6–8)*. Using an appropriate protocol, high levels of viability (>90%) may be observed *(7,8)* and viability levels are apparently unaffected by storage times up to 13 yr *(6)*. No single protocol has been found to be successful for a wide range of organisms. Direct immersion in liquid nitrogen, either with or without the addition of cryoprotectants, has been successfully employed to preserve a small number of unicellular Chlorococcales and cyanobacteria *(9,10)*. Most protocols that have been

From: *Methods in Molecular Biology, Vol. 38: Cryopreservation and Freeze-Drying Protocols*
Edited by: J. G. Day and M. R. McLellan Copyright © 1995 Humana Press Inc., Totowa, NJ

developed employ a simple two-step system with controlled/semicontrolled cooling from room temperature to an intermediate subzero holding temperature prior to plunging into liquid nitrogen.

The majority of protocols in current use have been developed empirically with the prefreezing culture regime, cryoprotectant choice and concentration, cooling rate, and thawing regime being manipulated to minimize the amount of damage to the algal cells and hence maximize postthaw viability levels. Damage may be induced by changes in the composition of intra- and extracellular solutions, mechanical and physical stresses caused by ice formation, as well as direct cold shock and chilling injuries.

High levels of viability are required to minimize the possibility of selecting an unrepresentative freeze–tolerant subpopulation. Furthermore, the relatively slow growth rate of algae (generation times in the range of 8–72 h are not uncommon), along with other problems associated with low viable cell density (*see* Note 1), can result in practical difficulties in reestablishing a viable culture. At the Culture Collection of Algae and Protozoa (CCAP), a minimum level of 50% postthaw viability is required before an algal strain is maintained only in a cryopreserved state. Currently, approx 30% of the collection is cryopreserved and another 5%, where viabilities are in the range of 15–50%, are maintained by both serial subculture and cryopreservation.

Only a relatively small taxonomic section of the algae currently held in culture collections has been successfully preserved. At present, 380 of the 470 strains cryopreserved at CCAP are members of the Chlorococcales or Cyanophyceae; most of these have been frozen using the protocol outlined below (*see* Section 3., steps 1–8).

2. Materials

1. Laminar flow cabinet and appropriate facilities for following good microbiological practice.
2. Culture: Use late log or early stationary phase cultures if available (*see* Note 2).
3. Medium: This is dependent on the nutritional requirements of the alga. For strains currently cryopreserved, the most commonly employed media are detailed in items 4–6.
4. BG 11 medium (Table 1), used for freshwater cyanobacteria and nonaxenic algae. Add 100 mL of solution 1, 10 mL of solutions 2–8, and 1 mL of solution 9 to distilled or deionized water to obtain a total vol of 1 L. pH is adjusted to 7.8 prior to sterilization.

Table 1
BG 11 Medium

Stock solutions	g/L
$NaNO_3$	15.0[a]
$K_2HPO_4 \cdot 3H_2O$	4.0
$MgSO_4 \cdot 7H_2O$	7.5
$CaCl_2 \cdot 2H_2O$	3.6
Citric acid	0.6
Ferric ammonium citrate	0.6
EDTA (disodium salt)	0.1
Na_2CO_3	2.0
Trace metal mixture:	
$\quad H_3BO_3$	2.86
$\quad MnCl_2 \cdot 4H_2O$	1.81
$\quad ZnSO_4 \cdot 7H_2O$	0.222
$\quad Na_2MoO_4 \cdot 2H_2O$	0.39
$\quad CuSO_4 \cdot 5H_2O$	0.079
$\quad Co(NO_3)_2 \cdot 6H_2O$	0.0494

[a]May be omitted for nitrogen fixing cyanobacteria.
Based on ref. *11*.

5. *Euglena gracilis:* Jaworski's medium (Table 2), used for axenic freshwater Chlorococcales. Add 0.5 mL of solutions 1–10 to distilled or deionized water to obtain a total vol of 1 L of medium. Then add 0.5 g sodium acetate (trihydrate), 0.5 g Lab-Lemco powder (Oxoid L29, Unipath Ltd., Basingstoke, UK), 1.0 g tryptone, and 1.0 g yeast extract.
6. Guillard's (f/2) medium (Table 3), used for marine algae. Add 10 mL of solution 1 and 1 mL of solutions 2–4 to filtered natural seawater to obtain a total vol of 1 L. Adjust pH to 8.0 prior to sterilization. Solution 5 (1 mL) is added to the medium prior to pH adjustment and sterilization if the medium is to be used for diatoms.
7. Cryoprotectant: 10% (v/v) Dimethyl sulfoxide (DMSO) in the appropriate medium. DMSO is cytotoxic and care should be taken when handling. Alternative cryoprotectants are also regularly employed for some strains (*see* Note 3).
8. Cryovials: Presterilized plastic screwcap 1.8 mL cryovials (Costar, Cambridge MA) (*see* Note 4).
9. Refrigeration system: Refrigerated methanol bath precooled to –30°C. **Note: Methanol is toxic and flammable.** Alternatively, more controlled rate coolers may also be used (*see* Note 5).
10. Liquid nitrogen dewar: Small 1–2 L wide-neck dewar.

Table 2
Euglena gracilis: Jaworski's Medium

Stock solutions	g/200 mL
$Ca(NO_3)_2 \cdot 4H_2O$	4.0
KH_2PO_4	2.48
$MgSO_4 \cdot 7H_2O$	10.0
$NaHCO_3$	3.18
$CaCl_2$	2.00
EDTA FeNa	0.45
EDTA Na_2	0.45
H_3BO_3	0.496
$MnCl_2 \cdot 4H_2O$	0.278
$(NH_4)_6Mo_7O_{24} \cdot 4H_2O$	0.2
Cyanocobalamin (vitamin B_{12})	0.008
Thiamine HCl (vitamin B_1)	0.008
Biotin	0.008
$NaNO_3$	16.0
$Na_2HPO_4 \cdot 12H_2O$	7.2

Based on ref. *12*.

Table 3
Guillard's (f/2) Medium

Stock solutions	g/L
$NaNO_3$	7.5
$NaH_2PO_4 \cdot 2H_2O$	5.65
Na_2EDTA	4.36
$FeCl_3 \cdot 6H_2O$	3.15
$CuSO_4 \cdot 5H_2O$	0.01
$ZnSO_4 \cdot 7H_2O$	0.022
$CoCl_2 \cdot 6H_2O$	0.01
$MnCl_2 \cdot 4H_2O$	0.18
$Na_2MoO_4 \cdot 2H_2O$	0.006
Cyanocobalamin (vitamin B_{12})	0.5[a]
Thiamine HCl (vitamin B_1)	100.0[a]
Biotin	0.5[a]
$Na_2SiO_3 \cdot 9H_2O$	40.18

[a]mg/l.
Based on ref. *12*.

11. Safety equipment: Long forceps, cryogloves, cryoapron, goggles.
12. Storage system: Cryogenic storage containers, with appropriate storage racks and inventory system suitable for holding cryovials (*see* Note 6).

13. Fluorescein diacetate (FDA) stain stock solution: 25 mg fluorescein diacetate in 25 mL of acetone.

3. Method

All culture manipulations should be carried out following good microbiological practice/aseptic technique, in a laminar flow cabinet if possible.

1. Grow cultures in the appropriate medium (*see* Section 2., items 3–6) under controlled environmental conditions. Flasks (50 mL) containing 30 mL of medium should be incubated static at 15°C under a photo fluence rate of 25–100 μmol m²/s (*see* Note 7). Light should be provided on a light:dark cycle, 16:8 h is generally regarded as optimal. Incubate cultures until they reach stationary phase; 30 d is used as a standard culture interval at CCAP.
2. Aseptically transfer sedimented cells with 15 mL of medium into a presterilized beaker. Alternatively, for uniform cell suspensions, centrifuge at 500*g* for 10 min then decant the supernatant and resuspend the algae in 15 mL of fresh sterile medium. Remove a 5-mL aliquot for use as a control for viability assays (*see* Section 3., step 11).
3. Add 10 mL of medium containing 10% (v/v) DMSO in culture medium to the remaining 10 mL of dense culture and mix thoroughly to give a final concentration of 5% (v/v) DMSO.
4. Dispense the mixture in 1-mL aliquots into sterile prelabeled cryovials. Seal the vials and incubate at room temperature for 5 min (*see* Note 8).
5. Transfer the filled cryovials to a precooled refrigerated bath (–30°C) and incubate for 15 min (*see* Note 9).
6. Transfer the vials rapidly, using forceps, to a wide-necked dewar containing liquid nitrogen. Then transport the vials in the dewar to the storage system.
7. Transfer the vials rapidly, using long forceps, to the storage/racking system. Storage is generally in the liquid phase of liquid nitrogen (*see* Note 6).
8. A full inventory/stocklist, including the locations of the vials within the storage system, should be maintained. This can most easily be retained either on a computer database, as card indices, or any other appropriate systems.
9. Thawing: Transfer stored cryovials (as in Section 3., step 6) to a dewar containing liquid nitrogen for transport and temporary storage. When ready, place the vials in a preheated water bath (40°C) and agitate until the last ice crystal has melted. On completion of the thawing remove immediately and transfer to the laminar flow cabinet (*see* Note 10).

10. Wipe the cryovial with 70% (v/v) ethanol to sterilize the outer surface of the vial. Aseptically transfer the contents into 30 mL of appropriate sterile medium and incubate as detailed in step 1 (*see* Note 11).
11. Viability assays:
    a. Vital staining: Add 50 µL of FDA stain stock solution to 1 mL of culture, incubate at room temperature for 5 min, and observe by blue light fluorescence microscopy (*see* Note 12). Viable cells fluoresce green (FDA positive) and nonviable cells appear red or colorless. It is necessary to dilute out cryoprotectant by transferring the thawed culture into 9 mL of appropriate sterile medium and incubating for up to 24 h prior to staining (*see* Note 13). Viability is expressed as a percentage of control (nontreated unfrozen culture; Section 2., item 2) vs FDA positive cells.
    b. Colony formation in agar: Dilute control cultures (Section 2., item 2) 1:1 with sterile medium. Transfer 1-mL aliquots of logarithmic dilutions of control and freeze–thawed cultures into sterile Petri dishes (50-mm diameter). Add approx 2.5 mL of 1.0% (w/v) agar in the appropriate medium at 40°C, agitate gently to mix culture, and agar. When the agar has gelled, seal the Petri dishes with Parafilm and incubate under standard growth conditions as detailed in step 1 (*see* Note 14). Colonies are counted using a dissecting microscope (magnification x50) and the viability of thawed cultures expressed as a percentage of control culture.

Where cultures form clumps or aggregates, colonies may arise from either an individual or clump of viable cells. Cellular multiplicity must be taken into account for both control and frozen and thawed cultures. The frequency of cell aggregation is determined by microscopic observation of at least 500 cells and the percentage postthaw viability determined as detailed:

Multiplicity	=	No. cells/No. clumps
Total cell number	=	No. colonies × multiplicity
Percentage viability	=	Total no. cells before treatment × 100/ Total no. cells after treatment

(*see* Note 16).

4. Notes

1. At low cell densities, photoinhibition may occur as a result of the lack of self-shading by other algal cells. This in turn may result in the death of cells that have survived cryopreservation. Where cultures are not axenic, low postthaw viabilities may result in the overgrowth of the alga by the competing associated microbial flora.
2. Stationary phase cultures are denser, thus reducing the time required to reestablish a viable culture. Also, stationary phase cultures may contain

intracellular storage products, including lipids, which may act as additional cryoprotectants.
3. DMSO may also be used at 10% (v/v) final concentration. Alternatively, glycerol at either 5 or 10% (v/v) or methanol at 5 or 10% (v/v) may be used. DMSO and glycerol are sterilized by autoclaving at double the final concentration in culture medium. Methanol is sterilized by filter sterilizing through an alcohol-stable 0.2-µm filter. Choice of cryoprotectant is largely dependent on its effectiveness and its cytotoxicity to the algal strain being frozen.
4. Costar vials have an internal rubber O-ring that appears to reduce the likelihood of liquid nitrogen leaking into the vial. Alternative cryovials, when used at CCAP, tend to have a significantly greater incidence of nitrogen leakage and subsequent potential contamination of algal cultures when stored under liquid nitrogen. Leakage can also lead to the possible rupturing of the vial on thawing, which has obvious safety implications (*see also* Note 10). Alternatively, Cryoflex (Nunc, Hastrup, Denmark) may be used. This will form a seal and prevent leakage into the vial.
5. Alternative methods of cooling to the holding temperature include the use of controlled-rate coolers. Slow cooling rates (1–5°C/min) are generally used for many larger and more complex microalgae *(13–15)*.
6. Vials must be stored at below –135°C. This is achieved by storage either in the liquid or vapor phase in an insulated cryobin or alternatively in a more sophisticated autofill cryostore. At CCAP, liquid phase storage is preferred as this ensures that stored material is always maintained at below –135°C.
7. Cyanobacteria, particularly when physiologically stressed, should be cultured at relatively low light levels. Light intensity is less critical for postthaw eukaryotic algae.
8. For some slow and nonpenetrating cryoprotectants (e.g., glycerol), cultures are incubated for 30 min in the presence of cryoprotectant prior to cooling.
9. For some algae, including members of the Prasinophyceae incubation for 30 min at –30°C is used (*see also* Note 5).
10. Liquid nitrogen leakage into the vial is potentially dangerous. On thawing the nitrogen will evaporate and this could cause the vial to rupture. Care should be taken when handling vials containing nitrogen and full safety equipment—gloves, apron, goggles, and so on—should be used. Liquid nitrogen contains low levels of viable bacteria and these in turn may contaminate axenic cultures if leakage occurs. Where alternative frozen specimens are available vials containing liquid nitrogen should be discarded.
11. For axenic cultures normally cultured on minimal mineral media, improved recovery rates may be obtained using medium with additional complex organic nitrogen, e.g., proteose peptone (Oxoid) at 1 g/L.

12. Temperature levels of the specimen increase rapidly during fluorescence microscopy and cells will die within 1–2 min. Observations and counts must be performed rapidly to prevent an underestimate of the viability level.
13. An incubation period that allows repair of sublethal damage but that is too short for cell division to occur, will give a more accurate index of viability when using vital staining rather than staining immediately after thawing.
14. Exposure time to the hot agar should be minimized; 40°C for more than a few minutes will be lethal for many nonthermophilic algae.
15. For most algae incubation at 25°C will reduce the period required to obtain discrete countable colonies.
16. For many algae, spreadplates of logarithmic dilutions may be used instead of pour-plates. Spread 0.1 mL of logarithmic dilutions of control and frozen and thawed culture onto the surface of Petri dishes (90 mm) containing an appropriate agar (1.5% w/v) solidified medium. Incubate and enumerate as detailed for pour-plates.

References

1. Day, J. G. (in press) Cryopreservation of microalgae and cyanobacteria, in *Proceedings of the Seventh International Congress of Culture Collections: Biodiversity and the Role of Culture Collections* (Fritz, D., ed.), WFCC, Braunschweig, Germany.
2. McGrath, M. S., Daggett, P., and Dilworth, S. (1978) Freeze–drying of algae: Chlorophyta and *Chrysophyta. J. Phycol.* **14,** 521–525.
3. Day, J. G., Priestley, I. M., and Codd, G. A. (1987) Storage, recovery and photosynthetic activities of immobilised algae, in *Plant and Animal Cells, Process Possibilities* (Webb, C. and Mavituna, F., eds.), Ellis Horwood, Chichester, UK, pp. 257–261.
4. Holm-Hansen, O. (1967) Factors affecting the viability of lyophilized algae. *Cryobiology* **4,** 17–23.
5. Holm-Hansen, O. (1973) Preservation by freezing and freeze–drying, in *Handbook of Phycological Methods: Culture Methods and Growth Measurements* (Stein, J., ed.), Cambridge University Press, Cambridge, UK, pp. 173–205.
6. McLellan, M. R. (1989) Cryopreservation of diatoms. *Diatom Res.* **4,** 301–318.
7. Morris, G. J. (1976) The cryopreservation of *Chlorella* 1. Interactions of rate of cooling, protective additive and warming rate. *Arch. Microbiol.* **107,** 57–62.
8. Morris, G. J. (1978) Cryopreservation of 250 strains of Chlorococcales by the method of two-step cooling. *Br. Phycol. J.* **13,** 15–24.
9. Holm-Hansen, O. (1963) Viability of blue-green algae after freezing. *Physiologia Planta.* **16,** 530–539.
10. Box, J. G. (1988) Cryopreservation of the blue-green alga *Microcystis aeruginosa. Br. Phycol. J.* **23,** 385,386.
11. Rippka, R., Dereuelles, J. B., Waterbury, J. B., and Herdman, M. (1979) Generic assignments, strain histories and properties of pure cultures of cyanobacteria. *J. Gen. Microbiol.* **111,** 1–61.

12. Thompson, A. S., Rhodes, J. C., and Pettman, I. (1988) *Culture Collection of Algae and Protozoa Catalogue of Strains.* Culture Collection of Algae and Protozoa, Ambleside, UK.
13. Morris, G. J., Coulson, G. E., and Engels, M. (1986) A cryomicroscopic study of *Cylindrocystis brebissonii* De Bary and two species of *Micrasterias* Ralfs (Conjugatophyceae, Chlorophyta) during freezing and thawing. *J. Expt. Bot.* **37,** 842–856.
14. Fenwick, C. and Day, J. G. (1992) Cryopreservation of *Tetraselmis suecica* cultured under different nutrients regimes. *J. Appl. Phycol.* **4,** 105–109.
15. Watanabe, M. M., Shimizu, A., and Satake, K. N. (1992) NIES-Microbial culture collection at The National Institute for Environmental Studies: cryopreservation and database of culture strains of algae, in *Proceedings of the Symposium on Culture Collection of Algae* (Watanabe, M. M., ed.), NIES, Tusukba, Japan, pp. 33–42.

CHAPTER 10

Cryopreservation of Plant Protoplasts

Brian W. W. Grout

1. Introduction

Isolated plant protoplasts are most commonly liberated from tissues, or cultured cells, by enzymic digestion of the retaining cell wall material *(1)*. This occurs in a solution hypertonic to the parent tissues that both induces plasmolysis and ruptures plasmodesmatal connections between cells so that spherical protoplasts, each representing the contents of a single cell, constitute the bulk of the released population. The hypertonic situation provides osmotic stability, compensating for the wall pressure of the original cells, and prevents protoplast lysis. Occasionally, mechanical disruption of tissue has been used to liberate protoplasts, although this produces a very low yield of protoplasts owing to the inefficiency of the process and the high probability of physical damage *(2)*. The isolated protoplasts will not be entirely representative of the parent tissues, as many cells, particularly senescent and highly vacuolate ones, are susceptible to the osmotic stresses and handling procedures involved in protoplast isolation and respond by rupturing. Such isolated protoplasts have a key role in pure and applied research and practical plant improvement, especially where they can be cultured in vitro to regenerate whole plantlets, and are being used for genetic manipulations of various types *(3–5)*. They are also invaluable for numerous more fundamental studies of structure and function where unencumbered access to the naked plasmalemma is desirable. This latter category includes the study of the biological effects of low temperatures and freezing conditions on plant material *(6–9)*.

Isolated protoplasts are essentially ephemeral structures, and if viable will regenerate wall-like components and structures at the plasmalemma surface, often within a few hours of their isolation *(1)*. As soon as this process begins, the essential and most valuable property of the protoplast, notably the accessability of the unencumbered plasmalemma surface, is lost. A technique for storage is, therefore, required that not only protects viability, but also suspends regenerative development, particularly the production and accumulation of materials at the protoplast surface. The ideal candidate as a conservation technology is cryopreservation, using liquid nitrogen (−196°C) as the storage medium, which will provide an effective cessation of all aspects of growth and development and provide the maximum available genetic stability *(10–14)*. At such a temperature, normal cellular chemical reactions do not occur as energy levels are too low to allow sufficient molecular motion to complete the reactions. Water exists either in a crystalline or glassy state under these conditions and has such a high viscosity ($>10^{13}$ poises) that rates of diffusion are insignificant over timespans measured at least as decades. The majority of the chemical changes that might occur in the protoplast are, therefore, effectively prevented and so it is stabilized to the maximum extent that is practically possible. Applications of cryopreservation might include conservation of protoplasts isolated from rare or unique parent material, storage of genetically manipulated material for subsequent use, storage to allow transport of protoplasts between laboratories, and as an experimental procedure in studies of plant cryobiology. Successful recovery of viable protoplasts from liquid nitrogen may be achieved either by relatively slow cooling-rate techniques, or by vitrification which is essentially independent of cooling rate. Slow cooling rates have been particularly effective as the route to successful cryopreservation of plant materials *(10–15)*. The theories and mechanisms of the process have been reviewed in detail elsewhere *(10–13,16,17)*, and the present discussion will be limited to an outline of the major principles and the practical difficulties that they generate. When protoplasts are frozen in a bulk volume of aqueous solution, at relatively slow cooling rates, ice will form first in the extracellular medium, which can be considered as a single large compartment when compared to individual cells. The probability of ice nucleation is related to the compartment size. The presence of extracellular ice, with continued cooling, will necessarily leave a residual, extracellular solution of decreasing volume and increasing hypertonicity (as it contains the bulk of the solutes of the original extracellular solution that are excluded from the ice crystal lattices) until the eutectic point is reached. A number of stresses will be generated in this situation, each of which is potentially damaging to the protoplasts. These include:

1. Reduction in temperature;
2. Mechanical effects of extracellular ice crystals;
3. Altered physical properties of residual solution, e.g., pH, viscosity;
4. Generation of gas bubbles and electrical fields at the ice–solution interface that can interact with the plasmalemma;
5. Concentration of extracellular solute with effects on water potential;
6. Volume reduction of the protoplast and consequent lesions of the plasmalemma evident on thawing;
7. Concentration of cytoplasmic solutes; and
8. Effects of a concentrating cytoplasm on intracellular compartments.

For successful preservation with slow cooling it is essential that, after the initial formation of extracellular ice, water is lost from the cells to the residual solution (cryodehydration) to build up the intracellular solute concentration. At this point the concept of optimal cooling rate becomes significant, as too rapid a rate will produce undercooling in the intracellular solutions and, therefore, a high probability of ice nucleation. If intracellular ice is generated it will be lethal. Similarly, too slow a cooling rate will reduce the rate of cryodehydration and extend the process time to the point where damaging effects will become significant. The intent of slow cooling protocols is to dehydrate the biological material as much as possible before reaching the eutectic temperature, while retaining viability with the aid of cryoprotectants, which are compounds that protect against various aspects of freezing injury *(10,17)*. At an appropriate subzero temperature, the sample is plunged directly into liquid nitrogen and subsequent survival will depend on vitrification of the intracellular solutions *(18,19)*. In addition to the difficulties of designing a suitable cryoprotectant regime, the rate of cooling before quenching has to be very precisely controlled, for the reasons just stated, which will require expensive, programmable equipment for the protocol design stage. Once a protocol is designed more inexpensive, rate-dedicated equipment might be applicable. Alternatively, vitrification as a cryopreservation technique relies on treatment of isolated protoplasts in the unfrozen state to manipulate the solute concentration in both extra- and intracellular solutions. The goal is to achieve a sufficiently high solute concentration to prevent the solution freezing into crystalline ice when cooled, and to ensure its transition into the amorphous, or glassy, state. To vitrify pure water requires an ultrarapid cooling rate and vitrification of any significant bulk of water is rarely obtained *(19,20)*. Increasing solute concentration makes vitrification easier, and reduces the required cooling rate to the point where simple quenching into liquid nitrogen will vitrify the solutions effectively *(18–20)*. Vitrification of isolated rye protoplasts *(21,22)* has been achieved using ethylene glycol solutions both as an internal cryoprotectant and dehydration protectant, and as an external dehydrating solution. Protoplasts in sample containers designed to present a low volume and high surface area can then be immersed directly into liquid nitrogen for storage without further regard to cooling rate. The survival of protoplasts immediately after thawing is simply estimated using a microscopical fluoroscein diacetate assay *(21)* that indicates the retention of active esterase enzyme within the intact protoplast plasmalemma. Viability is assessed by regrowth. Although the techniques of slow cooling and vitrification have both been successful in cryopreservation of protoplasts, no technique has yet emerged that has wide applicability in detail, and empirical optimization of procedures will be necessary in many instances. The practical details that follow refer to published, successful protocols and highlight the important and critical steps in the procedures. The information is intended to suggest a likely pattern of investigation for successful cryopreservation of new materials.

2. Materials

2.1. Slow Cooling

1. Protoplast cultures: These should be freshly isolated from the parent tissue to avoid the possibility of regeneration of extracellular material. Suspend the protoplasts in the appropriate growth medium for the system at a density of 10^4–10^6/mL. The use of parent plants that have been acclimated to low temperatures or cell cultures acclimated to osmotic stress may improve the survival of the isolated protoplasts following cryopreservation (*see* Note 1).
2. Culture medium: The basal medium must be one that has been proved as satisfactory for the maintenance and growth (if required) of the protoplast type under investigation. Add cryoprotectants, as required, to the complete growth medium. The basal medium will also be used for any washing steps and postthaw culture.
3. Cryoprotectants: A range of protectant compounds may be used that include dimethyl sulfoxide (DMSO), glycerol, sucrose, glucose, mannitol, and sorbitol. Analar-grade reagents should be used for cryoprotection and the

DMSO should be of spectroscopy-grade. DMSO should be stored above 19°C and not allowed to solidify. This material should not be stored for more than 9 mo in the laboratory if it is to be used as a cryoprotectant. Prepare the cryoprotectants (2x required concentration) in growth medium and adjust the pH to that of the protoplast incubation medium. The desired concentration is achieved on adding the cryoprotectant to the protoplast suspension. These solutions are best filter-sterilized, either by using disposable units attached to a hypodermic syringe or larger, vacuum-assisted filter systems. Care should be taken to ensure that the filter used is resistant to the solubilizing activity of DMSO when used at higher concentrations. Cryoprotectant solutions should be freshly made where possible but may be stored at 4°C for a maximum of 24 h. The most widely applicable cryoprotectant system for slow cooling-rate freezing of protoplasts has been 5% (v/v) DMSO + 10% (w/v) glucose *(23,24)*, but other cryoprotectant regimes have been successful *(11,23–27* and *see* Note 2). DMSO is cytotoxic and care should be taken in handling and dispensing material treated with DMSO.
4. Ampules: Single-use polypropylene ampules (2 mL) with a screwcap and sealing gasket are recommended. These can be marked with a water-resistant marker pen, but scoring of identification codes lightly into the body of the tube, using a heated needle, before freezing is recommended. Tubes can be purchased as presterilized or autoclaved in the conventional way.
5. Pipets: Solutions and protoplasts can be transferred using conventional, sterile Pasteur or volumetric pipets. Care must be taken to use resistant plastics when working with DMSO solutions at higher concentrations.
6. Freezing and storage apparatus: Slow cooling rates are most reliably achieved using purpose-built, programmable apparatus that is readily available from commercial sources. However, it is possible to construct a simple and reliable apparatus, albeit with a limited sample capacity, in the laboratory using conventional components *(12,28)*. Storage refrigerators are also widely available from commercial sources and should be liquid-nitrogen cooled, with a level-monitoring safety alarm. In the author's opinion, storage in the liquid, rather than vapor, phase is preferable and samples should be stored as multiple replicates and at different locations within the storage containers. Ideally, stored samples will be replicated between duplicate storage containers. A number of relatively shallow, wide-necked vacuum flasks are needed to hold small quantities of liquid nitrogen for various cooling procedures and for transferring frozen ampules to the storage refrigerator. Long (20 cm) metal forceps are required. Liquid nitrogen is an extremely hazardous compound and should not be used without full training and use of necessary protective clothing.

7. Thermostatted water bath at an appropriate temperature.
8. A freshly made stock solution of fluoroscein diacetate ($7.2 \times 10^{-3} M$) in acetone. The immediate survival of the protoplasts is determined by microscopical examination of a sample using UV light (excitation filter 355 nm; barrier filter 420 nm).

2.2. Vitrification

1. Protoplast cultures: Prepared in an appropriate isolation medium and then transferred to the suspending medium at a density of 3×10^6 protoplast/mL.
2. Suspending medium: A sorbitol solution ($0.53-1.03 M$ depending on acclimation state (*see* Note 2) + 1 mM CaCl$_2$, 1 mM Mes (pH 5.5) and 1% (w/v) bovine serum albumin (BSA).
3. Loading (equilibration) solution: A freshly made sorbitol solution isotonic with the chosen suspending medium + $1.5 M$ ethylene glycol (Analar) and 1% (w/v) BSA certified to 98% protein content (*see* Note 5).
4. Dehydration (vitrification) solution: A freshly made sorbitol solution ($0.88 M$) plus $7.0 M$ ethylene glycol and 6% (w/v) BSA (*see* Note 5).
5. Sample straws: These are 0.5 mL polypropylene straws commercially available for the preservation of animal semen. Unplugged straws that can be heat sealed at the ends, as required, are recommended.
6. Unloading solution: Details as described for the suspending medium but with the BSA content reduced to 0.2% (w/v) (*see* Note 5).

3. Methods

3.1. Slow Cooling

There is often considerable benefit to be gained from acclimation of the parent tissues for protoplast isolation to low growth temperatures *(21,22)*. The required treatment will need to be determined empirically for different species and varieties. The protocol outlined in step 1 is applicable to rye.

1. Move 1-wk-old plants from a 20/15°C, day–night temperature regime with a 16 h photoperiod to a 13/7°C regime and an 11.5 h photoperiod for 1 wk. Thereafter, hold the plants at a constant 2°C with a 10 h photoperiod for 4 wk prior to protoplast isolation.
2. Cool cryoprotectant solutions on ice and then aseptically add to a similarly cooled protoplast suspension in a stepwise fashion for over 30 min. Then incubate in the final concentration of protectant for another 30 min before cooling and freezing. When designing a protocol for a new situation, a protoplast sample should be assessed for survival and viability as appropriate following the cryprotectant pretreatment (*see* Note 3).

3. Transfer 1.0 mL of protoplast suspension to a polypropylene ampule and place in the freezing apparatus precooled to 4°C. A further sample should be assessed for survival/viability at this point, following removal of the cryoprotectant solutions (*see* step 6).
4. Cool the samples at 1 or 2°C/min to −35°C (*see* Note 6). Rapidly transfer the ampule directly into liquid nitrogen for prolonged storage. A sample should be thawed from −35°C (*see* step 6) and assayed for survival/viability.
5. When required, remove the frozen samples from the storage container and thaw by immediate immersion in water at 40°C with gentle agitation until ice is no longer visible in the sample tube.
6. Place the thawed samples on ice and the wipe the ampules with 70% (v/v) ethanol prior to opening under sterile conditions. Carefully transfer the protoplast suspension to a larger vessel and "dilute out" the cryoprotectants by at least a factor of 10 using stepwise addition of the original protoplast suspension medium over a 1-h period. Harvest the protoplasts and resuspended in an appropriate medium for subsequent regrowth or experimental use. Survival is determined as detailed in Section 2.1., item 8.

3.2. Vitrification

1. Acclimate parent plant material by growing under appropriate conditions (*see* Section 3.1. and Note 1).
2. Following isolation wash the protoplasts in culture medium, pellet, and resuspend in the suspending medium.
3. At the loading stage, add a 50-µL sample of protoplasts in suspending medium to 900 µL of loading medium, gently mix, and incubate at room temperature for 20 min.
4. Centrifuge the suspension from step 3 and discard the supernatant. Resuspend the pellet in the residual liquid after decanting and equilibrate to 2°C. Subsequently, add 900 µL of thermally equilibrated vitrification solution to the protoplast suspension and gently mix.
5. Draw 300 µL of the protoplast suspension into a polypropylene straw and heat seal both ends.
6. After a total of 60-s incubation in the vitrification solution, plunge the straw into liquid nitrogen, where it can remain for prolonged storage.
7. When required, remove the straw from the storage vessel. Thaw by holding in the air for 10 s and then incubate in ice/water for another 10 s. Cut both ends of the straw and expel the contents into ice-cold unloading solution. Rinse the straw with unloading solution.
8. After addition of the unloading solution mix the suspension and bring to 20°C for 15 min. Then pellet the protoplasts and resuspend in fresh unloading solution before their survival/viability is assessed (*see* Section 2.1., item 8; Note 3).

4. Notes

The procedures detailed in this chapter have been proven for a very limited range of plant types, and experience suggests details of the procedures may need to be changed when a new subject is investigated. The most productive areas for consideration are acclimation, cryoprotectant/ vitrification chemicals, and their concentration and cooling rates. The approach to modification of the procedures is essentially empirical and the following notes are intended to indicate the starting points for such investigations.

1. Acclimation: In plants acclimated to low temperatures, there are modifications to the lipid composition of the plasma membrane that reduce the extent of phase transition of membrane lipids from lamellar to the hexagonal phase during cryodehydration *(29,30)*. This removes the primary cause of the loss of osmotic responsiveness and destabilization of the plasma membrane observed in frozen-thawed protoplast suspensions of nonacclimated material. There is also significant accumulation of intracellular solutes with known cryoprotectant activity, such as glycine betaine, proline, and soluble sugars *(31)*. Attempts at acclimation of parent tissues should be made with regard to the physiology of the parent plant. If the parent tissue is a cell culture, it may be beneficial to attempt acclimation by osmotic stress; this has been successful in the cryopreservation of sycamore cell lines *(32)*.
2. Cryoprotectants: There is a wide range of compounds that have cryoprotectant activity and, in the case of new subjects, these must be borne in mind. The concentrations of protectant are critical to a successful protocol but the required concentrations and mixtures can vary widely between different plant types. There is no substitute for an empirical approach to this type of investigation and it is best started by exposure of the protoplasts to potential protectants for various incubation times, on ice, followed by assay for survival and viability. Thereafter the procedure can be repeated with the most promising materials in the presence of extracellular ice at a relatively high temperature, e.g., $-10°C$. The selection process is then developed by further temperature reduction.
3. Survival and viability: Survival of protoplasts following freeze–thaw procedures refers to their integrity and osmotic responsiveness and is determined by microscopic observation in the recovery medium and during limited hypotonic excursions, during which the protoplasts should be seen to exhibit Boyle van't Hoff behavior *(9)*. Survival can also be estimated by the fluoroscein diacetate technique described in the Section 3. Viability refers to the ability of the protoplasts to undergo their full developmental

cycle of growth and development, regenerating a cell wall and recovering their status as a cell.
4. Suspending medium: The sorbitol concentration of the suspending medium is likely to depend on the level of acclimation of the parent tissue, and the most suitable level will be empirically determined.
5. Vitrification solutions: It should be noted that these solutions have proven to be effective with rye protoplasts, but there is no assurance that they will be directly transferable to other subjects. Consequently, an investigation will need to made of compounds and concentrations in much the same way as for the cryoprotectants used in slow cooling procedures.
6. Slow cooling rates: The rate of cooling is essentially the way of controlling water loss from the protoplasts embedded in an ice–hypertonic solution matrix, and has been shown to vary widely in successful cryopreservation procedures for different plant tissues. A likely successful range is from 0.5–25.0°C/min, and the optimum rate can only be determined empirically. The use of a cryomicroscope to observe shrinkage rates during the course of the protocol is an invaluable aid in this process *(9)*.

References

1. Power, I. B. and Davey, M. (1990) Protoplasts of higher and lower plants, in *Methods in Molecular Biology 6: Plant Cell and Tissue Culture* (Pollard, J. W. and Walker, J. M., eds.), Humana Press, Clifton, NJ, pp. 237–259.
2. Cocking, E. C. (1972) Plant cell protoplasts—isolation and development. *Annu. Rev. Plant. Physiol.* **23**, 29–50.
3. Tomes. D. T. (1990) Current research in biotechnology with applications to plant breeding, in *Progress in Plant Cellular and Molecular Biology* (Nijkamp, H. J. J., Van Der Plas, L. H. W., and Van Aartrijk, J., eds.), Kluwar, The Netherlands, pp. 23–32.
4. John, M. E. and Stewart, J., McD. (1992) Genes for jeans: biotechnological advances in cotton. *Trends Biotechnol.* **10**, 165–169.
5. Pollard, J. W. and Walker, J. M. (eds.) (1990) *Methods in Molecular Biology 6: Plant Cell and Tissue Culture.* Humana Press, Clifton, NJ.
6. Steponkus, P. L. (1985) Cryobiology of isolated protoplasts—applications to plant cell cryopreservation, in *Cryopreservation of Plant Cells and Organs* (Kartha, K., ed.), CRC, Boca Raton, FL, pp. 49–60.
7. Steponkus, P. L. and Lynch, D. V. (1989) Freeze/thaw induced destabilisation of the plasma membrane and the effect of cold acclimation. *J. Bioenerg. Biomembr.* **21**, 21–41.
8. Langis, R. and Steponkus, P. L. (1991) Vitrification of isolated rye protoplasts: protection against dehydration injury by ethylene glycol. *Cryo-Lett.* **12**, 107–112.
9. Steponkus, P. L. and Weist, S. C. (1979) Freeze-thaw induced lesions in the plasma membrane in low temperature stress, in *Crop Plants—The Role of the Membrane* (Lyons, J. M., Douglas, G., and Raison, J. K., eds.), Academic, London, pp. 231–254.

10. Kartha, K. (ed.) (1985) *Cryopreservation of Plant Cells and Organs.* CRC, Boca Raton, FL.
11. Withers, L. A. (1987) The low temperature preservation of plant cell, tissue and organ cultures and seed for genetic conservation and improved agricultural practice, in *The Effects of Low Temperatures on Biological Systems* (Grout, B. W. W. and Morris, G. J, eds.), Edward Arnold, London, pp. 389–409.
12. Withers, L. A. (1990) Cryopreservation of plant cells, in *Methods in Molecular Biology 6: Plant Cell and Tissue Culture* (Pollard, J. W. and Walker, J. M., eds.), Humana Press, Clifton, NJ, pp. 39–48.
13. Grout, B. W. W. (1990) Genetic preservation in vitro, in *Progress in Plant Cellular and Molecular Biology* (Nijkamp, H. J. J., Van Der Plas, L. H. W., and Van Aartrijk, J., eds.), Kluwar, The Netherlands, pp. 13–22.
14. Grout, B. W. W. (1991) Cryopreservation of plant cells and organs, in *Conservation of Plant Genetic Resources Through In Vitro Methods* (Zakri, A. H., Normah, M. N., Senawi, M. T., and Abdul Karim, A. G., eds.), FRIM/MNCPGR, Kuala Lumpur, Malaysia, pp. 43–56.
15. Bajaj, Y. P. S. (1989) Cryopreservation of plant protoplasts, in *Biotechnology in Agriculture and Forestry 8: Plant Protoplasts and Genetic Engineering* (Bajaj, Y. P. S., ed.), Springer Verlag, Berlin, pp. 97–106.
16. Grout, B. W. W. and Morris, G. J. (1987) Freezing and cellular organisation, in *The Effects of Low Temperatures on Biological Systems* (Grout, B. W. W. and Morris, G. J., eds.), Edward Arnold, London, pp. 147–174.
17. Grout, B. W. W. (1991) The effects of ice formation during cryopreservation of clinical systems, in *Clinical Applications of Cryobiology* (Fuller, B. J. and Grout, B. W. W., eds.), CRC, Boca Raton, FL, pp. 81–94.
18. Fahy, G. M., MacFarlane, D. R., Angell, C. A. and Meryman, H. T. (1984) Vitrification as an approach to cryopreservation. *Cryobiology* **21**, 407–426.
19. Franks, F. (1985) *Biophysics and Biochemistry at Low Temperatures.* Cambridge University Press, Cambridge, UK.
20. Taylor, M. J. (1987) Physico-chemical principles in low temperature biology, in *The Effects of Low Temperatures on Biological Systems* (Grout, B. W. W. and Morris, G. J., eds.), Edward Arnold, London, pp. 3–71.
21. Langis, R. and Steponkus, P. L. (1990) Cryopreservation of rye protoplasts by vitrification. *Plant Physiol.* **92**, 666–671.
22. Langis, R. and Steponkus, P. L. (1991) Vitrification of isolated rye protoplasts: protection against dehydration injury by ethylene glycol. *Cryo-Lett.* **12**, 107–112.
23. Takeuchi, M., Matushima, H., and Sugawara, Y. (1982) Totipotency and viability of protoplasts after long-term freeze preservation, in *Plant Tissue Culture, 1982* (Fujiwara, A., ed.), Japanese Association for Plant Tissue Culture, Tokyo, pp. 797,798.
24. Withers, L. A. (1985) Cryopreservation of cultured plant cells and protoplasts, in *Cryopreservation of Plant Cells and Organs* (Kartha, K., ed.), CRC, Boca Raton, FL, pp. 243–267.
25. Bajaj, Y. P. S. (1988) Regeneration of plants from frozen (–196°C) protoplasts of *Atropa belladona* L., *Datura innoxia* Mill., and *Nicotiana tabacum* L. *Ind. J. Exp. Biol.* **26**, 289–292.

26. Gazeau, C. M., Hansz, J., Jondet, M., and Dereuddre, J. (1992) Cryomicroscopic comparison of freezing (to –40°C) and thawing effects on *Catharanthus* cells and their protoplasts. *Cryo-Lett.* **13,** 137–148.
27. Gazeau, C. M., Blanchon, C., and Dereuddre, J. (1992) Freeze-preservation of *Catharanthus* protoplasts at liquid nitrogen temperature. Comparison with cells. *Cryo-Lett.* **13,** 149–158.
28. Withers, L. A. and King, P. J. (1980) A simple freezing unit and cryopreservation method for plant cell suspensions. *Cryo-Lett.* **1,** 213–220.
29. Sugawara, Y. and Steponkus, P. L. (1990) Effect of cold acclimation and modification of the plasma membrane lipid composition on lamellar to hexagonal phase transitions in rye protoplasts. *Cryobiology* **27,** 667.
30. Webb, M. S. and Steponkus, P. L. (1990) Dehydration-induced hexagonal phase formation in phospholipid bilayers. *Cryobiology* **27,** 666,667.
31. Koster, K. L., Steponkus, P. L., and Lynch, V. (1989) Solute accumulation during cold acclimation of rye. *Plant. Physiol.* **89,** S-26.
32. Pritchard, H. W., Grout, B. W. W., and Short, K. C. (1986) Osmotic stress as a pregrowth procedure for cryopreservation 3. Cryobiology of sycamore and soybean cell suspensions. *Ann. Bot.* **57,** 379–388.

Chapter 11

A Two-Step or Equilibrium Freezing Procedure for the Cryopreservation of Plant Cell Suspensions

Elly W. M. Schrijnemakers and Frank Van Iren

1. Introduction

Plant cell and tissue culture is widely used in fundamental research (cell physiology, molecular biology and genetics, developmental biology, and so on) as well as in commercial activities (breeding and vegetative multiplication of food and feed crops and ornamentals). Several factors inherent to these types of work render the long-term storage of the materials urgent. Some of these are:

1. Costs of maintenance: Compared to the culture of many microorganisms, maintenance of plant cells and tissues is costly in terms of labor and equipment.
2. Stability: Plant cells have a remarkable flexibility, with respect to both composition of the genome (endoreduplication to higher ploidy levels is not uncommon in the plant kingdom; somatic mutations, aneuploidy, and amplification frequently occur in tissue cultures) and epigenetics (gene expression patterns are less stable than for most animal cell types). These mechanisms may render material with desired traits valueless. A well-recognized phenomenon is the loss during subculture of cells or tissues of the potential to regenerate into complete plants.
3. Patenting: As a rule, patented material has to be deposited in a publicly accessible culture collection.

In 1973 the first plant cell suspension was reported to survive cooling down to the temperature of liquid nitrogen (1). Since then many investigators have devised protocols for the cryopreservation of plant material.

For water-rich plant materials (including cell suspensions, small calli, embryos, apical meristems), most protocols were based on the two-step or equilibrium-freezing approach developed in the late 1970s *(2,3)*. Usually after some preculture of the cells, the method starts with the addition of cryoprotectants, which cause moderate dehydration. Severe dehydration of the specimen occrurs during the first freezing step resulting from the formation of ice crystals in the medium. During the rapid cooling in the second step, the contents of the shrunken cells should vitrify, i.e., form a glass that is stable below −120°C. Apparently, cells should attain concentrations of (suitable) solutes high enough to prevent ice crystal formation, but not high enough for the occurence of solute effects on cell constituents *(4)*. After cryostorage, the material is rewarmed rapidly because of the risk of crystal formation in the glass, followed by rapid and deleterious crystal growth *(5)*. Recently, a different approach to the cryopreservation of plant cells emerged *(6,7)*. This employs a vitrification procedure that avoids the formation of any ice crystal, within the cells as well as in the medium. The method is dealt with in Chapter 12.

We developed a standard protocol for plant cell suspensions (single cells and aggregates in liquid medium) based on a classical two-step approach *(8)*. During the past 6 yr this single protocol has proven to be successful with a large number of strains from a wide variety of species. The standard protocol has been applied to 11 species; rapidly growing cell lines were recovered from 8 of them, i.e., the monocots: rice *(Oryza sativa)*, barley *(Hordeum vulgare)*, and the dicots: tobacco *(Nicotiana tabacum)*, carrot *(Daucus carota)*, petunia *(P. hybrida)*, periwinkle *(Catharanthus roseus)*, and 2 species of *Tabernaemontana (divaricata* and *orientalis)*. The protocol failed for 3 other species, i.e., 2 species of *Cinchona (ledgeriana* and *robusta)* and one of *Tabernaemontana (longiflora)*. More than 60 strains have been tested, with 75% of them surviving our standard protocol.

The surviving lines differ widely; both morphologically and physiologically. Average cell diameters at the start of preculture vary from 15 (some rice strains) to 75 μm (one tobacco strain). Clusters of cells in the suspensions consist on the average of as little as 15 (many strains) to over 100 cells. Several had been genetically transformed (rice, barley, tobacco). Furthermore, growth rates of successfully preserved strains varied between a 10-fold increase in density in 1 wk to a 3-fold increase in 2 wk.

The nonsurviving strains include those that grow badly on solidified media, but several others were not apparently different from strains with high levels of postthaw survival. Some were sibling lines that had been cultured in parallel under slightly different conditions. Some of the strains did survive other protocols but a few are still recalcitrant to any of the procedures tested. In summary, our standard protocol appears to be widely appropriate; it was successful with over 70% of the strains to which it has been applied.

2. Materials

1. Programmable freezer able to cool at a rate of 1°C/min to –40°C. However, simpler means may suffice (*see* Note 1).
2. Liquid nitrogen storage vessel with drawers or canisters to accommodate cryovials.
3. Low speed benchtop centrifuge able to run at 100*g*, preferably with rapid acceleration and deceleration.
4. A device to accurately estimate the sedimented volume of the cells in a flask *(9)*.
5. A small orbital shaker in the vicinity of the flow cabinet.
6. Vials for freezing aliquots of suspension. Several companies sell cryovials (*see* Note 2).
7. Cell suspensions: These are cultured each in its own standard medium and incubated as follows: Temperatures range from 21–28°C, shaking velocities from 90–150 rpm on standard shakers, light regimes from dark to continuous light, subculture periods from 1–3 wk, inoculum dilutions between 3 and 10, closures of the flasks from tight aluminum foil to relatively permeable silicone foam stoppers.
8. Water: Use water of high purity (ultrafiltrated).
9. Chemicals at reagent grade: mannitol, sucrose, glycerol, fluorescein diacetate.
10. Activated charcoal at medicinal grade.
11. Petri dishes, 60-mm diameter sterile.
12. Whatman No. 1 filter paper disks, 42.5-mm diameter.
13. Widemouth (2 mm) sterile plastic Pasteur pipets.
14. DMSO (dimethyl sulfoxide): Purity should be checked at 275 nm. Absorbance should be below 0.03 *(10)*.
15. Agarose: As a rule, agarose gives better results than agar as a solidifier of the first recovery medium. Pronarose (Hispanagar, Burgos, Spain) may be successfully used as a cheap alternative.
16. Mannitol preculture medium: $1M$ (triple strength) in culture medium. Prepare the culture medium at double strength (in 50 mL) and add 18 g man-

nitol. Dissolve under gentle heating, add water up to 100 mL, and adjust pH. Autoclave 30 min at 110°C.
17. Cryoprotectant solution: SDG (sucrose, DMSO, glycerol) at double strength: $2M$ sucrose, $1M$ DMSO, $1M$ glycerol, in water (see Note 3). DMSO can be added before or after autoclaving for 30 min at 110°C (see Note 4). Store at 4°C for a maximum of 2 mo.
18. Recovery plates: Culture medium is supplemented with 0.75% (w/v) agarose and 0.5% (w/v) activated charcoal (Most strains appear to recover better in the presence of charcoal.) After autoclaving at 110°C, 6.5-mL aliquots are poured into 60-mm Petri dishes. Shortly before use, autoclaved filter paper disks are put on top of the medium.
19. Fluorescein diacetate (FDA) solution: 2 mg/mL fluorescein diacetate in acetone. Store in small volumes for maximum of 1 mo at –20°C (shorter periods if repeated warming). Approximately 10 µL is added to about 0.5 mL of cell suspension for viability assessment.
20. Safety equipment: Goggles, cryogloves, and so on.

3. Method

1. Use a cell culture when cell division rate is at its maximum and average cell volume is minimal (see Note 5). Allow the cells to settle. Substitute one-third of the volume (medium only) for mannitol preculture medium (mannitol concentration, $0.33M$). Incubate the cultures under their standard culture regime for an additional 3 d (see Note 6).
2. Place the culture on ice (flask tilted) and let the cells settle. Place the prepared cryoprotectant solution on ice.
3. Decant and discard all the medium above the cells. Measure the volume of the remaining suspension (see Section 2., item 4) and add the same volume of cold cryoprotectant (can easily be measured in a sterile disposable graded tube). Do not mix (see Notes 7 and 8).
4. Slowly shake the suspension, while on ice, on an orbital shaker at 80–100 rpm for 20–60 min.
5. Fill the cooled cryovials with 1.8 mL of cryoprotected cell suspension.
6. Shake ice and water off the rack with cryovials. Rapidly transfer the rack into the programmable freezer (see Note 1) precooled to 0°C, and immediately start the program, with 20–30 min incubation at 0°C.
7. Cool at 1°C/min to –35°C and maintain at –35°C for 30 min (see Note 9).
8. Rapidly transfer the frozen vials to a dewar containing liquid nitrogen. After 5 min the vials can be transferred to the drawers or holders of the liquid nitrogen storage vessel. The storage vessel may be filled with liquid nitrogen or, alternatively, the liquid nitrogen level may be maintained below the lowest vials (see Note 10).

9. After storage (*see* Note 11), transfer the vials to be thawed into a dewar flask. Release the cap of each vial briefly in the laminar flow cabinet to remove any liquid nitrogen (*see* Note 12).
10. Transfer the vial to a clean 40°C bath and gently turn it over and over until the last piece of ice disappears (*see* Notes 13 and 14). Each vial is immediately placed on ice.
11. Centrifuge the vials at 100g for 1 min, wipe the vials with ethanol (*see* Note 15), and remove the supernatant using a sterile Pasteur pipet (*see* Note 16).
12. Allow at least 15 min to elapse between steps 9 and 11. Transfer the cell suspension (packed cells) using a widemouth Pasteur pipet, without touching the rim of the vial, to a recovery plate. It is important to place the cells as close together as possible. The remaining cells, together with some of their supernatant (*see* step 11), can be used for viability staining (*see* step 16). Seal each plate with two layers of Parafilm and transfer them to their regular culture conditions (*see* Notes 17–19).
13. After 3 d, transfer the top filter to a fresh recovery dish.
14. When most of the filter has become covered with a thick layer of cells, transfer the cell mass to a fresh dish without filter paper (*see* Note 20).
15. When the latter has grown, the cell mass can be suspended in liquid medium.
16. After every step of the protocol a small sample should be taken for viability staining (*see* Section 2., item 19; Note 21). Prepare a slide (UV quality) for fluorescence microscopy with blue-UV excitation and a barrier filter >510 nm. The cytoplasm of viable cells fluoresces bright green, dead cells do not stain at all. As a rule, weakly fluorescing cells are considered to be nonviable (*see* Notes 21–23). It should be noted that this method does not give an absolute level of viability.

4. Notes

1. A programmable freezer is costly (usually over $15,000). The exact cooling curve of many specimens does not seem to be critical, therefore much less sophisticated (cheaper) alternatives have been used successfully for this first step, including a dip-cooler in a methanol bath *(8)*, a household freezer at –30°C *(11)*, devices hanging in the neck of a liquid nitrogen vessel, or even a styrofoam insulation box in a –80°C freezer *(12)*. Currently, we are studying the performance of the simple and cheap Nalgene Cryo 1°C Freezing Container "Mr Frosty" (Nalge Cy, Sevenoaks, UK). We filled its bath with ethanol (instead of isopropanol, and replaced it after every use) and placed the device, with 18 vials, at –80°C for 120–150 min before plunging the vials into liquid nitrogen. The average cooling rate within the vials over the first 40°C was close to 1°C/min when measured

using a thermocouple, slightly higher than with the programmable freezer (*see* Note 9). Of course, there is no plateau in the cooling (cf. Section 3., step 9), and the final temperature within the vials during this step comes close to –80°C. For the 15 cell lines tested to date, we found no difference in comparison with the standard procedure, but we cannot exclude the possiblilty that some strains are more sensitive to this cooling protocol.
2. Using 2-mL vials, more vials can be filled from the same batch and less storage space is occupied than with 5-mL vials. All vials tested accumulate liquid nitrogen during prolonged and immersed storage.
3. Some researchers prepare the cryoprotectant solution in culture medium.
4. Alternative cryoprotectant solution: SGProp. Propane-1,2-diol (Analar quality, BDH Laboratory Supplies, Poole, UK), a (rapidly) penetrating cryoprotectant may be used in the place of DMSO (same molar concentration). It is less toxic. Initial results with various strains indicate no difference.
5. In case of weekly subcultured strains we take 3- (usually optimal) to 4- (usually as satisfactory, but contains more cells) d-old cultures.
6. Instead of mannitol, several other substances have been used during preculture *(13)*. During this step, without which the method usually fails, the cells grow smaller and the central vacuole tends to split up in several smaller ones. Little is known about the relevance of the many other changes that occur within the cells *(14)*.
7. Some workers add the cryoprotectant solution gradually to the culture medium. We tried this but obtained no improvement. It should be noted that the solution is rather viscous and mixes gradually during the next step.
8. Alternative cryoprotectant cocktails may be employed including: various sugars, sugar alcohols, polyols, amino acids, DMSO, and/or glycerol *(15)*, with concentrations in the range of 1–2.5M. Some of these penetrate the cell membrane and are known to protect cell constituents (membranes, proteins) against the various cryostresses. Others do not penetrate but appear to protect by osmotically removing water from the cells and, probably, by influencing the cell surface.
9. Cryoprotection cocktails show undercooling to a variable, usually considerable, extent because crystallization is a stochastic process and because most cryoprotectants suppress crystallization. The latent heat of crystallization thus disturbs the cooling curve significantly and nonreproducibly. With SDG cryoprotectant we observed rises of temperature within replicate vials during the same freezing run from negligible up to 10°C. "Seeding" of the specimen may control this variability. This involves briefly interrupting the cooling program, after passage of the melting point of the solution (e.g., at –7°C), and the bringing the vials into contact with a very low temperature, either with liquid nitrogen cooled forceps *(16)* or by dip-

ping the tips of the vials into a cryogenic bath *(17)*. The local low temperature induces the first ice crystals and the cooling rate within the sample becomes less disturbed. Most commercial programmable freezers have a program for seeding, consisting of a full speed and short cooling step. It does not work with mixtures like SDG. As a result of crystallization heat, the average cooling rate between 0 and −30°C within the vial is usually approx 0.7°C (thermocouple measurements). We eliminated the supercooling phenomenon with the nucleating agent but found no significant improvement with normally surviving cell lines or with recalcitrant ones (*see also* Note 1).
10. In the latter case all vials are stored in the gas phase; no liquid nitrogen enters the vials and losses by evaporation are less. However, a serious disadvantage may be that the temperature for the uppermost vials may attain values over −110°C.
11. Storage time of up to 6 yr does not influence the viability of stored material. Therefore, the whole procedure can be tested with storage in liquid nitrogen of less than 1 h.
12. Care should be taken as liquid nitrogen evaporates rapidly and there is a potential risk of the vial rupturing; full safety precautions should be observed.
13. During immersion in the water bath, microorganisms may enter the rim of the screwcap of the vial. If infection rates become unacceptable, measures should be taken. These may include decontamination of the water bath and filling it with sterile water. Alternatively, place several vials at once in a Nalgene Floating Microtube Rack (cat. no. 5974, Nalgene Cy) and allow it to float in the water bath until only a small amount of ice remains, but the rewarming is slower.
14. Application of microwaves *(18)* theoretically results in a faster thawing rate. However, household microwave ovens operate at a frequency (2450 MHz) for which ice is almost transparent. In our laboratory, this resulted in no improvement relative to conventional rewarming.
15. Some bacteria can survive in 100% ethanol, 70% (v/v) should be used and batches replaced regularly.
16. Washing of the cells prior to plating, is practiced by some research groups *(17)*; this usually appears to be deleterious *(3)*.
17. Light-cultured suspensions should be covered with one or more sheets of filter paper, reducing the light intensity to approx 8 μmol/m^2/s.
18. For some strains, the results obtained by the standard protocol could be improved by small modifications, e.g., addition of an amino acid to the recovery medium *(19)*.
19. Recovery and commencement of growth of the cells after cryostorage takes longer than untreated suspensions, twice as long, as a rule. For example,

several of our strains (untreated) show macroscopically visible growth on standard recovery plates after 2 d. After cryopreservation using the standard procedure, these strains usually show macroscopic growth 3–5 d after plating.
20. Agarose may be replaced by agar at this stage of recovery.
21. Immediately after thawing, results may be variable; after about 15 min viable cells appear to have resealed their plasma membrane and FDA positive counts (*see* Section 2., item 19) are stable.
22. The percentage of cells reacting positively with FDA is usually >85% in untreated cultures and the preculture treatment does not appreciably lower this percentage. The staining level drops during cryoprotection by 10–30% and immediately after thawing staining levels from 50–60% are common. During the first 2–3 d after plating the figure usually drops further. So, the most crucial figure appears to be the level on the day of transfer. For success, 40–70% is normally required, although cultures may be recovered with levels of 10% FDA positive at this stage.
23. Damage to the cells appears inevitable. Therefore, it is important to know whether the properties of the culture have changed as result of the cryopreservation, e.g., as result of selection (the procedure killed a certain subpopulation), of altered gene expression pattern, or of genetic changes. There are few fundamental studies on this subject, but, in general, properties/characteristics of the cell cultures have been retained after cryopreservation *(3,17,19,20),* although this may not always be the case *(21).*

Acknowledgment

This study has been supported by the Ministry of Economic Affairs by means of a grant in the Innovation Oriented Program on Biotechnology to FVI.

References

1. Nag, K. K. and Street, H. E. (1973) Carrot embryogenesis from frozen cultured cells. *Nature* **245,** 270–272.
2. Withers, L. A. (1985) Cryopreservation of cultured plant cells and protoplasts, in *Cryopreservation of Plant Cells and Organs* (Kartha, K. K., ed.), CRC, Boca Raton, FL, pp. 243–267.
3. Withers, L. A. (1987) The low temperature preservation of plant cell, tissue and organ cultures and seed for genetic conservation and improved agricultural practice, in *The Effects of Low Temperatures on Biological Systems* (Grout, B. W. W. and Morris, G. J., eds.), Edward Arnold, London, pp. 389–409.
4. McLellan, M. R., Schrijnemakers, E. W. M., and Van Iren, F. (1990) The response of four cultured plant cell lines to freezing and thawing in the presence or absence of cryoprotectant mixtures. *Cryo-Lett.* **11,** 189–204.

5. Sakai, A. (1966) Survival of plant tissues at super-low temperatures. IV. Cell survival with rapid cooling and rewarming. *Plant Physiol.* **41,** 1050–1054.
6. Uragami, A., Sakai, A., Nagai, M., and Takahashi, T. (1989) Survival of cells and somatic embryos of *Asparagus officinalis* cryopreserved by vitrification. *Plant Cell Rep.* **8,** 418–421.
7. Langis, R. A., Schnabel, B., Earle, E. D., and Steponkus, P. (1989) Cryopreservation of *Brassica campestris L.* cell suspensions by vitrification. *Cryo-Lett.* **10,** 421–428.
8. Withers, L. A. and King, P. (1980) A simple freezing unit and routine cryopreservation method for plant cell cultures. *Cryo-Lett.* **1,** 213–220.
9. Blom, T. J. M., Kreis, W., Van Iren, F., and Libbenga, K. R. (1992) A noninvasive method for the routine-estimation of fresh weight of cells grown in batch suspension cultures. *Plant Cell Rep.* **11,** 146–149.
10. Matthes, G. and Hackensellner, H. A. (1981) Correlations between the purity of dimethyl sulfoxide and survival after freezing and thawing. *Cryo-Lett.* **2,** 389–392.
11. Sakai, A., Kobayashi, S., and Oiyama, I. (1991) Cryopreservation of nucellar cells of navel orange *(Citrus sinensis Osb.)* by a simple freezing method. *Plant Sci.* **74,** 243–248.
12. Maddox, A., Gonsalves, F., and Shields, R. (1982) Successful preservation of plant cell cultures at liquid nitrogen temperatures. *Plant Sci. Lett.* **28,** 157–162.
13. Göldner, E. M., Seitz, U., and Reinhard, E. (1991) Cryopreservation of *Digitalis lanata* cell cultures: preculture and freeze tolerance. *Plant Cell Tiss. Org. Cult.* **24,** 19–24.
14. Pritchard, H. W., Grout, B. W., and Short, K. C. (1986). Osmotic stress as a pregrowth procedure for cryopreservation. 1. Growth and ultrastructure of sycamore and soybean cell suspensions. *Ann. Bot.* **57,** 41–48.
15. Finkle, B. J., Zavala, M. E., and Ulrich, J. M. (1985) Cryoprotective compounds in the viable freezing of plant tissues, in *Cryopreservation of Plant Cells and Organs* (Kartha, K. K., ed.), CRC, Boca Raton, FL, pp. 75–113.
16. Dussert, S., Mauro, M. C., Deloire, A., Hamon, S., and Engelmann, F. (1991) Cryopreservation of grape embryogenic cell suspensions: 1. Influence of pretreatment, freezing, and thawing conditions. *Cryo-Lett.* **12,** 287–298.
17. Butenko, R. G., Popov, A. S., Volkova, L. A., Chernyak, N. D., and Nosov, A. M. (1984) Recovery of cell cultures and their biosynthetic capacity after storage of *Discorea deltoida* and *Panax ginseng* cells in liquid nitrogen. *Plant Sci. Lett.* **33,** 285–292.
18. Burdette, E. C., Wiggins, S., Brown, R., and Karow, A. M., Jr. (1980) Microwave thawing of frozen kidneys: a theoretically based experimentally-effective design. *Cryobiology* **17,** 393–402.
19. Meijer, E. G. M., Van Iren, F., Schrijnemakers, E., Hensgens, L. A. M., Van Zijderveld, M., and Schilperoort R. A. (1991) Retention of the capacity to produce plants from protoplasts in cryopreserved cell lines of rice *(Oryza sativa). Plant Cell Rep.* **10,** 170–177.
20. Rueff, I., Seitz, U., Ulrich, B., and Reinhard, E. (1988) Cryopreservation of *Coleus blumei* suspension and callus cultures. *J. Plant Physiol.* **133,** 414–418.
21. Mannonen, L., Toivonen, L., and Kauppinen, V. (1990) Effect of long term preservation on growth and productivity of *Panax ginseng* and *Catharanthus roseus* cell suspensions. *Plant Cell Rep.* **9,** 173–177.

Chapter 12

Vitrification of Plant Cell Suspensions

Poula J. Reinhoud, Atsuko Uragami, Akira Sakai, and Frank Van Iren

1. Introduction

Deposition at cryogenic temperatures is generally considered to be the safest method for long-term storage of living materials. (For a general introduction into the storage of plant cell cultures, refer to Chapter 11.)

Formation of ice crystals within the cell appears to be deleterious for almost any cell. Therefore, cryopreservation strategies should aim at the formation of a glass within the cells. Until recently, most plant cell and tissue cultures have been cryopreserved according to one of the "two-step" or equilibrium-freezing protocols. These start with a moderate dehydration of the cells in a mixture of cryoprotectants at a concentration of 1–2M. During slow cooling (typically at a rate of 1°C/min, down to –35°C), water in the medium freezes, and the gradually increasing concentration of cryoprotectants causes increasing dehydration of the cells. Some cryoprotectants also penetrate into the cells where they apparently protect against various cold, freezing, and dehydration related damage. At the end of the first step, cells should be approximately in osmotic equilibrium with the fluid phase of the medium. The second step is rapid cooling by immersion of the vials in liquid nitrogen, when the cell contents should solidify without crystallization, i.e., vitrify.

In 1985 the first successful cryopreservation of animal cells employing the so-called "vitrification" method was reported *(1)*. The basic feature of the method is the rapid (one-step) cooling of the pretreated cells in liquid nitrogen, resulting in glass formation in the extracellular medium as well as in the cells. Several years later the first plant cells, tissues, and organs were successfully "vitrified" *(2–4;* for a review, *see* ref. *5).* All these results were achieved using mixtures of one or two classical protectants, such as glycerol, dimethyl sulfoxide (DMSO), sorbitol, or polyethylene glycol (PEG) and the low-mol wt diols, 1,2-ethanediol (ethylene glycol), and 1,2-propanediol (propylene glycol). The vitrification solution itself should contain a very high concentration of suitable solutes in which no ice crystals are formed during cooling or rewarming. Such solutes should inhibit crystallization during cooling from the melting point downward. In addition, they should significantly elevate the glass transition temperature below which the solution does not crystallize. The faster the cooling and rewarming, the lower the probability of crystal formation in the (shortened) dangerous trajectory between the melting point and the glass transition. Successful vitrification solutions have total concentration of solutes in the range of 5–8M. They cause severe cellular dehydration.

Part of the components in the cocktail should penetrate into the cells for internal cryoprotection and vitrification (as in the two-step method). However, at full strength, all present vitrification solutions appear toxic to plant cells. Therefore, the usual strategy is not to allow the cells to equilibrate with them. In some protocols the cells first equilibrate with a lower concentration of cryoprotectants (loading) and the final amount is added just prior to rapid cooling to –196°C. In others, the cocktail is added in one step, but care is taken that rapid cooling commences before the cells have taken up lethal amounts of cryoprotectant or have become excessively dehydrated. Depending on the permeability of the cell membranes to the protectants and the sensitivity of the cells, penetration can be fine-tuned by altering the duration and/or temperature of exposure to the cryoprotectant cocktail. On thawing, the cryoprotectant should always be rapidly diluted with nontoxic substance(s), to prevent cell damage and death.

When comparing the methods of equilibrium-freezing and vitrification, it is clear that, for the latter method, the ultimate dehydration of the cells and additional intracellular cryoprotection can be regulated more or less independently by manipulation of the composition of the loading and vitrification solutions, as well as their application times and temperatures. Both addition of the cryoprotectant and dehydration can be achieved relatively quickly and at suprazero temperatures using this method. In the equilibrium-freezing technique, final dehydration and additional cryoprotection are attained coupled, gradually, and at subzero temperatures. Some potential additional advantages of the vitrification method should be mentioned. Where the two-step freezing is successful, ice crystals in the extracellular medium appear harmless. However, recalcitrant cells may be damaged by them. They are avoided in the vitrification method. Furthermore, the rapidity of cooling in the vitrification method may (partially) circumvent potential dangers including cold denaturation of proteins and phase segregations in biomembranes. Once a cocktail is found that vitrifies, and does not kill too many cells during the pretreatments, the additional damage from the excursion to liquid nitrogen may be small. We usually find the drop in viability resulting from the actual cooling and rewarming of cryoprotected cells to be much less in the vitrification method than in two-step freezing. A practical point concerns the absence of expensive controlled cooling equipment in vitrification. However, this can be avoided in two-step freezing as well (*see* Chapter 11).

Details of two vitrification protocols (a rapid one for robust strains and an elaborate one, including preculture and pretreatments, for strains that appeared more easily damaged by vitrification) are described.

2. Materials

1. Translucent 0.5-mL cryogenic straws: Several companies sell 13-cm long straws, one end of these straws has a plug with powder which becomes gas-tight on wetting (*see* Note 1). Cryovials can be used as an alternative (*see* Note 2).
2. Sealing powder for straws: eight different colors can be obtained from IMV (Instruments de Medicine Veterinaire, l'Aigle, France) (for alternatives, *see* Note 3).
3. Sterile, transparent, calibrated 10-mL tubes.
4. Bucket-type dewar: 2 L, in order to provide enough working space for rapid immersion of straws.
5. Liquid nitrogen storage vessel with drawers or canisters to accommodate straws and/or cryovials. Colored goblets facilitate identification of straws during retrieval from the collection.
6. Low-speed bench centrifuge able to run at $100g$ with rapid acceleration and deceleration.
7. 60-mm Diameter sterile plastic Petri dishes.
8. Whatman (Maidstone, UK) No. 1 filter paper disks, 42.5-mm diameter.
9. Presterilized glass Pasteur pipets, long size.
10. Widemouth sterile plastic Pasteur pipets.

11. Water: As a routine we use water of high purity (ultrafiltrated). Influence of singly distilled or demineralized water has not been investigated.
12. Chemicals at reagent grade: Mannitol, sucrose, glycerol, 1,2 ethane diol.
13. DMSO: Purity should be checked at 275 nm. Absorbance should be below 0.03 *(6)*.
14. Agarose: As a rule, agarose gives better results than agar when used as a solidifier of the first recovery medium.
15. Mannitol preculture medium: $1M$ mannitol (triple strength) in culture medium. Because of the large volume of the mannitol, 100 mL of the medium can be made as follows. Prepare regular culture medium at double strength, add 18-g mannitol to 50 mL of medium. Dissolve under gentle heating add water up to 100 mL, and adjust pH to that of the regular culture medium. Autoclave for 20 min at 120°C.
16. Vitrification solution PVS2 (*see* Note 4): $3.26M$ glycerol, $2.42M$ 1,2-ethane diol, $1.9M$ DMSO, prepared in water (or culture medium) with $0.4M$ sucrose. Adjust pH and autoclave for 20 min at 120°C. The solution can be stored at 4°C for at least 2 mo.
17. Pretreatment solution PVS_{20}: One part PVS2 plus four parts $0.4M$ sucrose in water (or culture medium). Adjust pH and autoclave for 20 min at 120°C.
18. Washing solution: $1.2M$ sucrose in water. Adjust pH and autoclave for 20 min at 120°C.
19. Recovery plates: Culture medium supplemented with 0.75% (w/v) agarose. After autoclaving (20 min, 120°C) it is poured into 60-mm diameter sterile plastic Petri dishes (6 mL/dish). Before use, two autoclaved filter paper disks are placed on top of the medium.

3. Methods
3.1. Elaborate Procedure (see Note 5)

1. Use a suspension culture when cell division rate is at its maximum and average cell volume is minimal. Allow the cells to settle. Decant 30% of the volume (medium only) and replace with mannitol preculture medium, giving a mannitol concentration of $0.33M$. Incubate the cultures under their regular culture regime for another 3 d.
2. Centrifuge for 1 min at $100g$ in a conical tube an amount of suspension that yields a pellet of 0.7–1 mL of packed cells. Discard the supernatant.
3. Make up to 5 mL with cold PVS_{20} (4°C), mix gently, and place on ice.
4. After 5 min, add five batches of 1 mL of cold full strength PVS2 at 1-min intervals.
5. Centrifuge for 1 min at $100g$ and discard the supernatant.
6. Make up to 1 mL with cold PVS2 within 2 min after the last addition (*see* step 4). Subsequently, add four batches of 1 mL of PVS2 at 1-min intervals.

7. As quickly as possible, fill six or more (*see* Note 6) straws with a sterile glass Pasteur pipet (≈0.5 mL each) by inserting the tip deeply into the straw, wetting the powder in the closure, and retracting and squeezing in such a way that stretches of suspension and small air bubbles alternate. (For rapid cooling in cryovials as an alternative, *see* Note 2.)
8. Quickly seal each straw by drumming sealing powder several millimeters into the open end. Make sure that the powder is wetted by the suspension in the straw.
9. Quickly immerse each straw in liquid nitrogen (*see* Note 7) for a few seconds, and leave it. Within another few seconds the straws sink and are ready for transfer to the storage containers.
10. After storage, warm one straw by gripping it securely (*see* Note 7) at one end and rapidly immersing and moving it for about 3 s in a clean water bath at 40°C (*see* Note 8). As soon as the contents are thawed, i.e., have become completely transparent (*see* Note 9), remove the straw.
11. Sterilize the surface of the straw with ethanol (*see* Note 10).
12. Cut the lower end of the straw with a sharp, and carefully sterilized, pair of scissors. Hold the straw over a tube with 7-mL cold washing solution. Cut the upper end and dispense the contents of the straws into the washing solution without touching the tube with the straw. Mix the cell suspension (depending on the sensitivity of the cells for full-strength PVS, mixing can be postponed). Empty five additional straws into the washing solution, mix, and leave for 20 min. Six straws should yield approx 2 mL of suspension, with approx 1 mL remaining in the straws.
13. Centrifuge for at least 1 min at 100g. If cells do not settle, even when centrifuged at 150g for 3 min, add 1 mL of standard culture solution prior to centrifuging in order to lower the density of the suspending liquid. Remove the supernatant (but retain it for the next step).
14. Transfer the cell mass using a wide mouth plastic Pasteur pipet from the tube to a recovery plate. Place the cells on the smallest possible area, i.e., close to each other. (The cells remaining in the tube can be used for viability staining; some of their supernatant [*see* step 13] should be added, to wash them from the tube.)
15. After 2 d, transfer the top filter to a fresh recovery dish.
16. When the layer of cells has become several millimeters thick, transfer the cell mass to a fresh dish without filter paper. (Agarose is, as a rule, no longer necessary—agar can be used instead.)
17. When the cell culture on the plate grows rapidly, a liquid suspension culture may be reestablished. Sometimes, addititional subculture on agar is necessary.
18. Viability may be monitored throughout the procedure by means of vital staining or regrowth tests (*see* Chapter 11).

3.2. Rapid Procedure (see Note 11)

1. Start with a rapidly growing cell culture. Transfer approx 2 mL to a 10-mL conical tube and allow the cells to settle. The volume of the cells should be approx 0.2 mL.
2. Discard the supernatant, add 4 mL of PVS2, and mix.
3. Immediately centrifuge for 20 or 30 sec at 100g, discard the supernatant, and add 2 mL of PVS2, either at room temperature or prechilled, and mix. In the latter case, chill the whole tube.
4. Incubate at room temperature for 1–7 min or on ice for 3–20 min.
5. Fill four straws as detailed in Section 3.1., step 7.
6. Follow the protocol detailed in Section 3.1., steps 8–18. However, four straws instead of six may be sufficient to initiate a new culture.

4. Notes

1. We use type 101 (cattle, nonsterilized) cyrogenic straws from IMV.
2. As an alternative, cryovials may be used. When filled with the amount used in straws, vitrification of PVS2 is usually attained. This is not withstanding the fact that the cooling rate is below 300°C/min in the vials *(7)* against approx 2000°C/min in 0.5 mL straws *(8)*. Obvious disadvantages include: devitrification is not easily visible, and that vials take more storage space.
3. Alternatively, seal with a hot pair of forceps or a sealing machine. When sealed properly, penetration of liquid nitrogen during prolonged storage can be minimized (*see* Note 9).
4. PVS2 stands for plant vitrification solution 2, and has been used for several types of plant cells and tissues *(7–9)*.
5. The elaborate procedure appeared to be successful (unpublished observations) for several cell suspension lines. One strain of each of the species: *Catharanthus roseus* (periwinkle), *Nicotiana tabacum*, *Daucus carota* (carrot), and *Hordeum vulgare* (barley) survived equally well or better using this procedure (described in Section 3.2.) as from two-step freezing (as described in Chapter 11). Where tested, they appeared to require preculture and failed to survive rapid vitrification. One tobacco and one *Catharanthus* line, both recalcitrant to two-step freezing, survived this protocol, although the latter required elaborate preculture. One, highly recalcitrant line of *Cinchona robusta* did not survive any treatment tested up to now.
6. If six straws are filled, the remaining 2 mL can be used for viability/regrowth tests. Apart from a negligible amount required for vital staining, this material, washed and plated for a regrowth test, may be used as the treated, unfrozen control. Note that the total time between the last addition of PVS2

(in step 6) and step 9 is crucial; the highest concentration of PVS being highly toxic.
7. Special straw-adapted forceps can be used, but long, robust laboratory pincers are equally effective.
8. Unacceptable contamination rates may be owing to this step. Sterilization of the water bath, and/or filling it with sterile water immediately before use and/or placing it in the laminar flow cabinet, may reduce the incidence of contamination.
9. During the warming of the straws, the contents should be carefully observed. It is possible that evaporation of liquid nitrogen, which has penetrated the straw, will expel the powder plug at one end together with some of the cell suspension. Care should be taken and appropriate safety procedures followed. Appearance of a bright and white lane, running through the straw and persisting for several seconds, indicates poor devitrification with rapidly proceeding recrystallization. This may be owing to a relatively large amount of cells, which diluted the vitrification solution excessively (pure vitrification solution should never show this phenomenon). In many straws a transient haze is observed. This probably results from (rapidly redissolving) microscopic gas bubbles. This is observed in straws in which cells survive, therefore the phenomenon appears harmless.
10. Some bacteria have been demonstrated to survive in ethanol. Replace regularly.
11. The rapid method has been successfully applied to embryogenic cell suspensions from various citrus species *(7,9)*. However, results were negative with several other (nonembryogenic) suspensions. For example, for some *Catharanthus*, rice, and tobacco lines, the viability levels, measured by vital staining, were low and regrowth was absent. A new rapid protocol, including a new vitrification solution and a loading step, but without preculture *(10)* is presently being tested for its applicability on a wider range of cell cultures.

Acknowledgment

This study has been supported by the Dutch Ministry of Economic Affairs by means of a grant in the Innovation Oriented Program on Biotechnology to FVI.

References

1. Rall, W. F. and Fahy, G. M. (1985) Ice-free cryopreservation of mouse embryos at –196°C by vitrification. *Nature* **313**, 573–575.
2. Uragami, A. Sakai, A., Nagai, M., and Takahashi, T. (1989) Survival of cells and somatic embryos of *Asparagus officinalis* cryopreserved by vitrification. *Plant Cell Rep.* **8**, 418–421.

3. Langis, R. A., Schnabel, B., Earle, E. D., and Steponkus, P. (1989) Cryopreservation of *Brassica campestris L.* cell suspensions by vitrification. *Cryo-Lett.* **10,** 421–428.
4. Towill, L. E. (1990) Cryopreservation of isolated mint shoot tips. *Plant Cell Rep.* **9,** 178–180.
5. Steponkus, P. L., Langis, R., and Fujikawa, S. (1992) Cryopreservation of plant tissues by vitrification, in *Advances in Low Temperature Biology,* vol. 2 (Steponkus, P. L., ed.), JAI Press, London, pp. 1–62.
6. Matthes, G. and Hackensellner, H. A. (1981) Correlations between the purity of dimethyl sulfoxide and survival after freezing and thawing. *Cryo-Lett.* **2,** 389–392.
7. Sakai, A., Kobayashi, S., and Oiyama, I. (1990) Cryopreservation of nucellar cells of navel orange *(Citrus sinensis Osb. var. brasiliensis* Tanaka) by vitrification. *Plant Cell Rep.* **9,** 30–33.
8. Niino, T., Sakai, A., Yakuwa, H., and Nojiri, K. (1992) Cryopreservation of *in vitro*-grown shoot tips of apple and pear by vitrification. *Plant Cell Tiss. Org. Cult.* **28,** 261–266.
9. Sakai, A., Kobayashi, S., and Oiyama, I. (1991) Survival by vitrification of nucellar cells of navel orange (*Citrus sinensis* var. *brasiliensis* Tanaka) cooled to –196°C. *J. Plant Physiol.* **137,** 465–470.
10. Nishizuwa, S., Sakai, A., Amano, Y., and Matsuzawa, T. (1993) Cryopreservation of asparagus (*Asparagus officinalis*) embryogenic suspension cells and subsequent plant regeneration. *Plant Sci.* **91,** 67–73.

CHAPTER 13

Cryopreservation of Shoot-Tips and Meristems

Erica E. Benson

1. Introduction

The ability to regenerate whole plants from cryopreserved, meristematic shoot tissues provides a useful method for conserving plant genetic resources. This technique is especially important for vegetatively propagated species, or, for plants that produce recalcitrant seeds. Within this chapter, methodology is presented for those vegetative, meristematic tissues (apical and axillary shoot-tips, nodes, and buds) that have the potential to develop new shoots and regenerate whole plants after cryogenic storage. The term "shoot meristem" is frequently used erroneously in the context of cryopreservation. The shoot meristem is anatomically defined as a structure that contains the apical dome, and the youngest, unexpanded leaf primordia directly associated with the dome meristem. In practice, it is the larger shoot-tip used in "meristem cryopreservation." This structure comprises the meristem apex, subjacent tissue, and several larger, and often expanded, leaf primordia. Shoot-tip size and origin (e.g., apical or axillary) are critical factors in influencing the ability of tissues to survive cryopreservation. It is also possible to freeze nodal stem cuttings containing an axillary meristem and the dormant, and/or cold hardened buds of woody perennial species. Many species have the ability to survive shoot-tip cryopreservation, however, the routine application of one protocol to a range of species is limited by lack of reproducibility (e.g., in terms of regeneration potential). This may be owing to cellular heterogeneity of shoot-tips, inadequate tissue culture regimes, genotype, and physiological differences *(1–5)*. Initially, progress in the development of cryopreservation protocols for shoot-tips was similar to that of plant cell suspensions. Early methods involved the application of single or combined mixtures of chemical cryoprotectants and controlled cooling of tissues to intermediate freezing temperatures followed by transfer to liquid nitrogen (for an example, *see* ref. *4*). Under conditions of controlled cooling, a water vapor deficit is created (between the inside and outside of the cell) when extracellular ice is formed. Thus, intracellular water moves across the plasmalemma and cellular dehydration results. Under optimum conditions this is advantageous as the amount of water available for intracellular ice formation is reduced. However, it is also important to consider that excessive dehydration can lead to the toxic concentration of solutes and cryoprotectants can be harmful to the cell *(6,7)*.

Recent developments (encapsulation, cryoprotective dehydration, and vitrification) in plant tissue cryopreservation offer alternative and promizing methods for cryogenic storage that do not involve the precise control of cooling and freezing rates *(8–11)*. A major advantage of these techniques is that tissues may be plunged directly into liquid nitrogen, circumventing the need for expensive programmable freezing equipment. However, these methods will require optimizing for a particular plant system. Minor modifications to a protocol (e.g., duration and temperature of exposure to vitrification solutions, level of tissue dehydration) may significantly enhance postthaw survival. Thus, the development of successful cryogenic storage protocols depends on the ability to combine optimized component steps (pregrowth, cryoprotection, freezing, thawing, and recovery) in such a way as to maximize postfreeze regeneration.

At present there is no one method of shoot-tip cryopreservation that may be applied to a range of plant systems. However, there are several approaches that have been shown to be successful for a wide range of species. The methodology sections of this chapter are therefore presented as four basic protocol options that offer the possibility of being applied to diverse plant systems.

2. Materials

Where appropriate, materials should be supplied sterile and all manipulations performed in a Laminar flow bench. It is assumed that adequate safety equipment (face shield, protective gloves) will be used during all liquid nitrogen handling procedures.

2.1. Shoot-Tip Dissection

1. Binocular, dissecting microscope (magnification x20).
2. 2 × 10-mL syringes fitted with hypodermic needles.
3. Liquid growth medium (*see* Note 1).
4. 50-mm Petri dish lined with a filter paper.
5. Scissors, scalpel, forceps, Pasteur pipets.

2.2. Materials Common to All Cryopreservation Options

1. Liquid nitrogen.
2. Small benchtop dewar of 1-L capacity.
3. Cryovials, canes, Pasteur pipets, 90- and 50-mm Petri dishes, sterile filter papers, forceps, scissors, scalpels, heated magnetic stirrer, binocular dissecting microscope.
4. Long-term storage dewar with an appropriate drawer or cane inventory system.
5. Recovery media (*see* Note 2).
6. Safety equipment: cryogloves, goggles, and so on.

2.3. Materials Specific to Each Cryopreservation Protocol

2.3.1. Chemical Cryoprotection and Controlled Freezing

1. 100 mL of 10 and 5% (v/v) solutions of spectral grade, dimethyl sulfoxide (DMSO) made up in liquid medium (*see* Notes 3–5).
2. A programmable freezer.
3. A water bath set at 45°C.

2.3.2. Cryoprotective Dehydration

1. One or more of a series of pregrowth media containing $0.5–1M$ of a dehydrating agent (selected from sucrose, mannitol, or sorbitol) made up in standard solid medium (*see* Note 6).
2. An airtight vessel containing activated silica gel.

2.3.3. Vitrification

1. 100 mL of PVS2 vitrification solution *(9,11)* comprising standard liquid growth medium containing $0.4M$ sucrose to which is added: 30% (v/v) glycerol, 15% (v/v) ethylene glycol, 15% (v/v) DMSO (*see* Notes 3–5).
2. A dilution series, range (50–80% [v/v]) of PVS2 solution made in liquid medium (*see* Note 6).
3. Ice and ice bucket.

4. 100 mL Unloading solution comprising 1.2M sucrose made up in liquid culture medium.
5. A water bath set at 45°C.

2.3.4. Encapsulation and Dehydration

1. A 5-mL Gilson Pipetman fitted with a tip (*see* Note 7).
2. Liquid culture medium containing 0.75M sucrose (*see* Note 6) aliquoted (20 mL) into 100-mL conical flasks.
3. Calcium-free, liquid culture medium containing 3% (w/v) sodium alginate (2% low viscosity, as the sodium salt derived from sea Kelp) dispensed in 20-mL aliquots into McCartney bottles (*see* Note 8).
4. Liquid culture medium containing 100-mM $CaCl_2$ dispensed as 30-mL aliquots into conical flasks or beakers.
5. A 250–500 µm mesh sieve.
6. Sterile tissues or pieces of filter paper.

3. Methods

3.1. Tissue Preparation for Noncold Acclimated Species

As appropriate, select meristematic shoot tissues (nodal stem cuttings, apical shoot-tips, axillary shoot-tips, or axillary shoots stimulated by apex removal from field, glasshouse, or in vitro cultivated plants [*see* Note 9]).

3.2. Tissue Preparation for Cold Acclimated Species

Select cold-hardened dormant buds that have undergone natural seasonal acclimation. Alternatively, induce cold hardiness in rejuvenated buds or *in vitro* plants using temperature-controlled environments (*see* Note 10). If appropriate, investigate the effects of low temperatures (–5 to 10°C), treatment duration (days to months), and temperature cycling (diurnal alternation of growth and low temperature treatments) on the ability of shoot-tips to survive cryopreservation *(8,12–15)*. The gradual reduction of prefreeze temperatures over several days before shoot-tip dissection offers another possibility (*see* Note 10).

3.3. Shoot-Tip Dissection

1. With the aid of a dissecting microscope and two hypodermic needles, remove the larger expanded leaves from the shoot-tips, while retaining several non- or partly expanded leaf primordia and a small amount of subjacent tissue.
2. Trim the shoot-tip to an approximate size of 1–5 mm (*see* Note 11).
3. Transfer shoot-tips to a filter paper soaked with pregrowth medium immediately after dissection (*see* Note 12).

3.4. Chemical Cryoprotection and Controlled Freezing

1. Transfer dissected shoot-tips from filter paper bridges to solid or liquid medium containing 5% (v/v) DMSO (*see* Notes 3–6).
2. Pregrow for 1 d (*see* Notes 6 and 13).
3. Transfer 10–20 shoot-tips to a cryovial containing 0.5 mL 10% (v/v) DMSO made up in standard liquid culture medium (*see* Notes 3–6).
4. Incubate at room temperature for 1 h (*see* Note 6).
5. Transfer to a programmable freezer and set a freezing cycle (*see* Note 14) with parameters in the following ranges:
    a. Cooling rate (–0.25 to –5°C/min);
    b. Terminal transfer temperature (0 to –40°C); and
    c. Holding time at temperature of transfer (30–45 min).
6. On completing the controlled temperature program, immediately transfer the vials to a small dewar filled with liquid nitrogen located by the side of the freezer. Transfer vials to a large storage dewar as appropriate (*see* Note 15).
7. For thawing immediately after removal from the long-term storage dewar, transfer the vials to a 45°C water bath. Once ice has melted transfer the vials to a flow bench area.
8. Wipe the outside of the vial with a tissue to remove any water. Expel the contents of the vial onto filter papers contained in a Petri dish, the cryoprotectant is removed by capillary action. Transfer the shoot-tips to recovery medium (*see* Note 2).

3.5. Cryoprotective Dehydration

This method is adapted from those of Niino et al. *(8)*, Uragami *(9)*, and Bagniol and Engelmann *(16)*.

1. Remove nodal segments or shoot-tips (2–5 mm) from in vitro plantlets, trim away any expanded leaves with a scalpel (*see* Note 16).
2. Culture the nodes or shoot-tips on a series of media containing 0.5–0.75M of dehydrating agent (e.g., sucrose, mannitol, or sorbitol in standard solid medium) for 1–5 d (*see* Note 6).
3. Transfer plant tissues to standard medium and determine posttreatment survival and regeneration, and select the regime that maximizes dehydration with minimal loss of viability (*see* Note 17).
4. After step 3 has been optimized, transfer the dehydrated nodes to airtight vessels containing activated silica gel, and desiccate for 2–16 h.
5. Place nodes on recovery medium and determine which treatment correlates maximum dehydration with minimal loss of viability.
6. Choose the optimal pregrowth treatment and test the efficacy of this treatment in permitting postfreeze recovery after ultrarapid freezing as follows.

7. Transfer the dehydrated shoot-tips to cryovials and plunge directly into liquid nitrogen.
8. On thawing, remove the vials from the dewar and thaw at ambient temperatures.
9. Transfer tissues to recovery medium (*see* Notes 2 and 18).

3.6. Vitrification

This method is based on that of Uragami *(9)* and Sakai et al. *(11)*. For further details of this procedure *see* Chapter 12.

3.6.1. Preparation of Stock PVS2 Solution and Supplementary Solutions

1. Prepare liquid growth medium containing $0.4M$ sucrose.
2. Add: 30% (v/v) glycerol, 15% (v/v) ethylene glycol, 15% (v/v) DMSO (*see* Note 19).
3. Prepare a range of diluted PVS2 solutions (e.g., 10–80% [v/v] PVS2 solution made up in standard liquid medium) and a solution of standard liquid medium containing $1.2M$ sucrose (*see* Note 20).

3.6.2. Vitrification Procedure

This initially involves a toxicity trial (to determine the optimum method for adding the vitrification solution) followed by a cryopreservation procedure.

1. Prepare shoot-tips (Section 3.3.), place in a cryovial, and construct a toxicity test for the PVS2 solutions as follows.
2. Add a 1-mL aliquot of PVS2 in the concentration range of 50–80% (v/v), expose the tissues to a specific concentration for 1–10 min, remove almost all the vitrification solution from the tissues, and replace with a step-wise, higher concentration of PVS2 solution (*see* Note 6).
3. Gradually increase the concentration of PVS2 until the tissues are finally exposed to the 100% solution (*see* Note 21).
4. Remove the 100% vitrification solution and replace with unloading solution containing $1.2M$ sucrose in standard liquid medium. Perform 2 to 3 washes in fresh unloading solution and maintain in this solution for up to 30 min.
5. Expel the content of the vials onto a filter paper and allow the paper to soak away excess vitrification solution.
6. Transfer the shoots to recovery medium and determine the survival and regeneration characteristics after the different treatments. Select the treatment that permits maximum survival and maximum cryoprotection.

7. On exposing to 100% PVS2 solution (after optimization), transfer the vials directly to liquid nitrogen.
8. For rewarming: Place in a water bath at 45°C and recover shoot-tips as described in steps 5 and 6.

3.7. Encapsulation and Dehydration

This novel approach to shoot-tip cryopreservation was developed by Fabre and Dereuddre *(10)* who applied it for the first time to *Solanum phureja*.

1. Transfer freshly dissected shoot-tips (Section 3.3.) to a 3% (v/v) alginate solution; swirl gently to submerge the shoots. Avoid agitation which will form air bubbles.
2. Using a 5-mL Pipetman, withdraw 3 mL of alginate solution at the same time, capture 2–4 shoot-tips.
3. Hold the pipet vertically and wipe away any alginate solution from the exterior of the tip using sterile filter papers (*see* Note 22).
4. Place the pipet vertically over the calcium containing medium and with slow and controlled pressure dispense droplets of the alginate into the liquid. On contact with Ca^{2+} the droplets will solidify and form a gel encapsulating the shoot-tips.
5. When all the shoot-tips are encapsulated, remove the excess liquid and dispense the beads into a large Petri dish. Sort out the beads, which contain the shoot-tips, transfer to a dish, and leave to polymerize for 30 min.
6. Transfer the beads to liquid dehydration medium containing $0.75M$ sucrose made up in liquid medium. Transfer to a reciprocal shaker platform and leave for 3 d (*see* Note 6).
7. Retrieve the beads from the sucrose solution by sieving and blot them dry on filter papers. Transfer beads to an open, empty Petri dish located at the back of a horizontal filter type flow bench. Desiccate the beads for 4 h (*see* Note 23).
8. Transfer beads to a cryovial and plunge directly into liquid nitrogen storage (*see* Note 24).
9. On thawing, remove the vials and rewarm at room temperature.
10. Expel the content of the vials in to a clean Petri dish and transfer to recovery medium.

3.8. Postfreeze Recovery Assessments
3.8.1. Criteria for Evaluating Successful Shoot-Tip Cryopreservation

It is not possible to perform viability assays based on microscopy (e.g., vital staining) on shoot tissues that are too optically dense. A general procedure for assessing morphogenic development during short-term recovery is given:

1. Within 7 d of thawing, examine tissues under a binocular microscope.
2. Record the number of shoot-tips that show greening and expansion of the leaf primordia (*see* Note 25).
3. Continue to perform weekly assessments and record a time course of recovery events as leaf expansion, callus formation, and shoot and plantlet regeneration (*see* Note 26).
4. Evaluate recovery as normal events (plantlet formation in the absence of callogenesis and adventitious shooting) and abnormal events (callus and adventitious shoot development) (*see* Notes 27 and 28).

4. Notes

1. Tissue culture media are highly species specific. It is assumed that the investigator will already have optimized tissue culture protocols for the material to be cryopreserved.
2. The composition of the recovery medium will be species specific. However, there is frequently the requirement to supplement recovery media with growth regulators to support shoot regeneration.
3. Cryoprotectants and pregrowth additives must be of high purity; spectroscopically pure DMSO is recommended.
4. Cryoprotectants and vitrification solutions are usually prepared in standard liquid culture media. It is advisable to check the pH of the medium after incorporating the additives.
5. Where possible, cryoprotectant mixtures should be sterilized by filtration. However, this approach is unsuitable for high viscosity vitrification solutions, for these additives, autoclaving is the only practical method of sterilization.
6. The concentration, duration, and temperature of exposure to chemical additives are critical factors in developing cryoprotective dehydrating treatments. The range of parameters included in this chapter serve only as guide. For some systems, it may be necessary to investigate treatment concentrations and exposure periods with those recommended.
7. To ensure uniformity of bead size, and, hence, dehydration characteristics, it is important to use an alginate-dispensing aperture of a constant size. For larger plant structures use a pipet tip with the tip removed using a hot scalpel blade.
8. Prepare calcium-free standard liquid medium (Ca^{2+} salts of vitamins accepted) and add 3% (w/v) sodium alginate (2% low viscosity kelp alginate as the sodium salt). It is very difficult to solubilize alginate and this is best achieved by placing the liquid medium on a magnetic stirrer, heat to boiling while agitating vigorously, and add small amounts of alginate, step by step. Discard flocculated solutions.

9. These tissues may exhibit different responses to freezing. Prefreeze factors, known to influence survival include:
   a. Culture duration;
   b. Pre- and postfreeze light regime;
   c. Location of axillary shoot relative to apical shoot;
   d. Duration of subculture interval before freezing; and
   e. Developmental stage and number of leaf primordia associated with the apical dome.
10. This is especially important for woody perennial species; seasonal screening for freeze tolerance is required to determine at which period of the growth cycle the buds are best able to survive cryopreservation *(12–15)*. Experimentally induced cold hardiness is frequently used to enhance recovery in cryopreserved shoot-tips derived from tissue cultures.
11. It is important to standardize shoot-tip and nodal cutting size, particularly as tissue size may affect the rate of air dehydration.
12. Shoot-tips rapidly dehydrate after dissection and they should be transferred to a suitable medium immediately after removal.
13. Newly dissected tissues are frequently sensitive to freezing injury and a period of pregrowth, often in the presence of chemical additives, can significantly enhance recovery *(4,7,16–19)*. Pregrowth treatments do not provide adequate protection against exposure to liquid nitrogen and additional cryoprotection is also required. A pregrowth strategy is developed by comparing the efficacy of several treatments. Pregrowth agents are incorporated into solid media or in filter papers moistened with additive-containing liquid media. DMSO is one of the most commonly used pregrowth additives. It is usually applied at lower concentrations than required for cryoprotection *(1)*. Other additives include: sorbitol, mannitol, sucrose and abscisic acid *(16–19)*.
14. Modern programmable freezers consist of a freezing chamber cooled by liquid nitrogen. With the use of temperature probes (connected to the sample and the chamber) and precise computer programming, it is possible to investigate a wide range of cooling/freezing parameters. With an appropriate output device (e.g., chart recorder) the temperature at which extracellular ice is formed (latent heat of crystallization) may be determined, with reasonable accuracy. This event, termed nucleation, can mark the onset of intracellular dehydration and it may be an important factor in developing a controlled cooling method for shoot-tips. Extracellular ice formation causes a vapor pressure deficit that is compensated for by the movement of intracellular water to the outside of the cell. The effect is "cryoprotective" as the amount of intracellular water available for ice formation is reduced. In the absence of external intervention, it is possible for

extracellular ice nucleation to occur randomly. However, this may be problematic and lead to variable freezing responses. The control of nucleation is thus an important consideration in developing slow freezing methods for shoot-tips. Some programmable freezers are fitted with a device that initiates nucleation by mechanically agitating the cryogenic samples. It is also possible to induce nucleation by touching the outside of the tube with a liquid nitrogen-chilled instrument.

15. It is important that vials are plunged immediately into liquid nitrogen after they are removed from the programmable freezer. Rewarming of the samples during this transition is detrimental to the tissues. For convenience a small dewar (e.g., of 1 L capacity) can be located near the programmable freezer and used to transport vials to the long-term storage dewar site.
16. Nodal segment stem size may be difficult to standardize.
17. Determination of water loss (using controlled oven drying) aids in the development of cryoprotective dehydration protocols.
18. To reduce osmotic shock, thawed tissues may be sequentially transferred to a range of media containing decreasing concentrations of a nontoxic osmoticum, such as sucrose.
19. The mixture will be extremely viscous and takes some time to completely dissolve in solution. Gradual addition of the cryoprotectants during vigorous agitation on a magnetic stirrer is recommended.
20. In some cases, shoot-tips may not survive direct exposure to 100% PVS2 and the stepwise addition of PVS2 in an increasing concentration series is necessary. Following vitrification and exposure to liquid nitrogen the shoots are initially recovered in an unloading solution ($1.2M$ sucrose). This prevents osmotic shock while replacing the toxic vitrification solution with a less damaging sucrose solution.
21. Gradually add increasing concentrations of PVS2 solution within the concentration range 50–100% (v/v). Exposure duration is short (5–30 min), and will require species-dependent determination. As an example, duration of exposure to 100% PVS2 is usually within the range of 5–20 min. Toxic effects of vitrification cocktails may be reduced by adding the solutions at chilling temperatures on ice. Thus, it may be possible to expose tissues to higher concentrations of PVS2 if the solutions are applied at 0°C.
22. Failure to do this will result in nonuniform beads. Alginate beads become extremely hard when air-dehydrated, and are sometimes difficult to manipulate. They rehydrate within an hour of placing on solid medium. Encapsulated shoot-tips may be thawed at ambient temperatures and placed on culture medium. Regenerating shoots grow out of the bead, although not all species respond in this manner and it may be necessary to remove the alginate, to allow regeneration to proceed.

23. Another option is to desiccate the beads over activated silica gel.
24. In some cases survival capacity may be increased by performing a controlled cooling cycle (*see* Section 3.4., step 5), before exposing the encapsulated shoot-tips to liquid nitrogen.
25. This is usually the first postfreeze recovery response.
26. Shoot-tip cryopreservation can only be considered successful if whole plants are regenerated from the frozen meristem. Postthaw recovery events include several morphological patterns of development (leaf expansion, callus, and root formation), which indicate postthaw viability, but not regeneration capability. Ideally, regeneration should proceed via an original meristem (primary or axillary) and not an indirect adventitious route (e.g., organogenesis from a callus). This criterion is important to ensure the genetic stability of plants regenerated from cryopreserved tissues. Quantitatively acceptable levels of recovery are of obvious importance. In the first stages of method development, survival levels approaching 50% of the total number of shoot-tips frozen may be considered encouraging. The application of the technique on a reproducible and routine basis demands high levels of regeneration (e.g., approx 80% of the total number of shoots frozen).
27. To reduce risks of somaclonal variation, avoid postfreeze regeneration via a dedifferentiated route.
28. For detailed studies (outside the scope of this chapter) concerning postfreeze morphogenetic and genetic evaluations, *see* Harding *(20),* Harding and Benson *(21),* Benson et al. *(22),* and Ward et al. *(23).*

References

1. Harding, K., Benson, E. E., and Smith, H. (1991) The effects of tissue culture duration on post-freeze survival of shoot-tips of *Solanum tuberosum. Cryo-Lett.* **12,** 17–22.
2. Benson, E. E., Harding, K., and Smith, H. (1989) Variation in recovery of cryopreserved shoot-tips of *Solanum tuberosum* exposed to different pre- and post-freeze light regimes. *Cryo-Lett.* **10,** 323–344.
3. Dereuddre, J., Fabre, J., and Bassaglia, C. (1988) Resistance to freezing in liquid nitrogen of carnation (*Dianthus caryophyllus* L. var Eolo) apical and axillary shoot-tips excised from different age *in vitro* plantlets. *Plant Cell Rep.* **7,** 170–173.
4. Henshaw, G. G., O'Hara, J. F., and Stamp, J. A. (1985) Cryopreservation of potato meristems, in *Cryopreservation of Plant Cells and Organs* (Kartha, K. K., ed.), CRC, Boca Raton, FL, pp. 159–170.
5. Manzhulin, A. V. (1983) Factors affecting the survival of potato stem apices after deepfreezing. *Fiziologiya Rastenii* **31,** 639–645.
6. Finkle, B. J., Zavala, M. E., and Ulrich, J. M. (1985) Cryoprotective compounds in the viable freezing of plant tissues, in *Cryopreservation of Plant Cells and Organs* (Kartha, K. K., ed.), CRC, Boca Raton, FL, pp. 75–113.

7. Kartha, K. K. (1985) Meristem culture and germplasm preservation, in *Cryopreservation of Plant Cells and Organs* (Kartha, K. K., ed.), CRC, Boca Raton, FL, pp. 115–134.
8. Niino, T., Sakai, A., and Yakuwa, H. (1992) Cryopreservation of dried shoot tips of Mulberry winter buds and subsequent plant regeneration. *Cryo-Lett.* **13,** 51–58.
9. Uragami, U. (1991) Cryopreservation of Asparagus (*Asparagus officinalis* L.) cultured in vitro. *Res. Bull. Hooaido, Natl. Agric. Exp. Stn.* **156,** 1–37.
10. Fabre, J. and Dereuddre, J. (1990) Encapsulation-dehydration: a new approach to cryopreservation of Solanum shoot-tips. *Cryo-Lett.* **11,** 413–426.
11. Sakai, A., Kobayashi, S., and Oiyama, I. (1990) Cryopreservation of nucellar cells of navel orange (*Citrus sinensis* Osb. var. *brasiliensis* Tanaka) by vitrification. *Plant Cell Rep.* **9,** 30–33
12. Tyler, N. and Stushnoff, C. (1988) Dehydration of dormant apple buds at different stages of cold acclimation to induce cryopreservability in different cultivars. *Can. J. Plant Sci.* **68,** 1169–1176.
13. Kuoksa, T. and Hohtola, A. (1991) Freeze-preservation of buds from Scots pine trees. *Plant Cell Tiss. Org. Cult.* **27,** 89–93.
14. Reed, B. M. (1988) Cold acclimation as a method to improve survival of cryopreserved *Rubus* meristems. *Cryo-Lett.* **9,** 166–171.
15. Reed, B. M. (1989) The effect of cold hardening and cooling rate on the survival of apical meristems of *Vaccinium* species frozen in liquid nitrogen. *Cryo-Lett.* **10,** 315–322.
16. Bagniol, S. and Engelmann, F. (1991) Effects of pregrowth and freezing conditions on the resistance of meristems of date palm (*Phoenix dactylifera* L. var. *Bou Sthammi Noir*) to freezing in liquid nitrogen. *Cryo-Lett.* **12,** 279–286.
17. Kartha, K. K. (1982) Cryopreservation of germplasm using meristem and tissue culture, in *Application of Plant Cell and Tissue Culture to Agriculture and Forestry* (Tomes, D. T., Ellis, B. E., Kasha, K. J., and Peterson, R.L., eds.), University of Guelph Publishers, Guelph, Canada, pp. 139–161.
18. Reed, B. M. (1993) Responses to ABA and cold acclimation are genotype dependent for cryopreserved blackberry and raspberry meristems. *Cryobiology* **30,** 179–184.
19. Kartha, K. K., Leung, N. L., and Gamborg, O. L. (1979) Freeze-preservation of pea meristems in liquid nitrogen and subsequent plant regeneration. *Plant Sci. Lett.* **15,** 7–15.
20. Harding, K. (1991) Molecular stability of the ribosomal RNA genes in *Solanum tuberosum* plants recovered from slow growth and cryopreservation. *Euphytica* **55,** 141–146.
21. Harding, K. and Benson, E. E. (1994) A study of growth, flowering and tuberisation in plants derived from cryopreserved potato shoot-tips: implications for *in vitro* germplasm collections. *Cryo-Lett.* **15,** 59–66.
22. Benson, E. E., Lynch, P. T., and Jones, J. (1992) The detection of lipid peroxidation products in cryoprotected and frozen rice cells: Consequences for post-thaw survival. *Plant Sci.* **85,** 107–114.
23. Ward, A. C. W., Benson, E. E., Blackhall, N. W., Cooper-Bland, S., Powell, W., Power, J. B., and Davey, M. R. (1993) Flow cytometric assessments of ploidy stability in cryopreserved dihaploid *Solanum tuberosum* and wild *Solanum* species. *Cryo-Lett.* **14,** 145–152.

CHAPTER 14

Cryopreservation of Seeds

Hugh W. Pritchard

1. Introduction

Storage of seeds is arguably the most effective and efficient method for the *ex situ* preservation of plant genetic resources. Low storage costs, combined with ease of seed distribution and regeneration of whole plants from genetically diverse material, offer distinct advantages for the storage for conservation of seeds compared with other types of plant tissues, such as meristems and pollen.

Conventional seed gene banks maintain seeds at about 5% (w/w) moisture content (fresh weight basis) and −18°C. Under such storage conditions it is predicted that high levels of viability will be retained over many decades *(1)*. The possibility that much longer periods of seed storage are attainable at ultra-low temperatures using cryopreservation techniques has been advocated, on the basis that biochemical processes are so reduced in cryopreserved material that biological deterioration is virtually stopped *(2,3)*. However, evidence of improved physiological preservation of seeds when cryopreserved compared to when stored at conventional gene bank temperatures is limited, for example, the loss in seed viability of an orchid species is lower at −196°C than at −20°C *(4)*. Nonetheless, it is possible to estimate the theoretical benefits of adopting cryopreservation for seed storage by extrapolation from published rates of viability loss at higher temperatures *(1)*. Based on the combined analysis of longevity data for the seeds of eight species stored at 90 to −13°C *(1)*, it is projected that the logarithm of seed longevity could be 2.25 U greater if liquid nitrogen rather than conventional freezer storage is used. This means that the standard deviation of the lifespan of individual seeds in a population stored under the same moisture conditions could be about 175 times greater; and in viability terms, the time for viability to fall by one on the probit scale (e.g., from 97.7–84.1%) could be 175 times longer. Thus, the relative benefits to seed longevity of adopting cryopreservation as the storage system may be considerable. It is important to note, however, that such theoretical considerations are made on two tentative assumptions. First, it is assumed that an exponential function can be used to describe the effect of temperature on the rate of seed viability loss. The alternative temperature model fitted to the data *(1)* includes a quadratic function that predicts that the logarithm of longevity of lettuce seeds, for example, stored at 5% moisture content would be the same (i.e., 0.2 d) at −150°C as at 80°C. Moreover, longevity at −196°C is theoretically shorter than this, although there is no evidence for this in practice, even for moist seeds *(5,5a)*. Second, it is supposed that the effects of temperature and moisture content on seed viability loss are independent *(1)*, although evidence for the existence of such a relationship down to cryopreservation temperatures is not yet available.

Accepting that there are likely to be some benefits to seed longevity through the use of cryopreservation, it is pertinent to consider which types of species are most suited to preservation in this way. Ideal candidates might be species with inherently short-lived seeds and endangered species with critically small population sizes; both types of material would benefit from the unlimited storage potential that liquid nitrogen storage appears to offer. However, it is not possible at this stage to recommend the use of cryopreservation over conventional seed bank storage for the conservation of a majority of desiccation tolerant, or "orthodox," seeds; the practical benefits of the latter method currently appear to outweigh the potential longevity gains associated with the former.

Early indications that seeds could tolerate cryopreservation treatment are referred to more fully in a comprehensive review of the subject *(3)*. Interestingly, work in this laboratory nearly 100 yr ago showed that seeds of 12 species from 8 families survived 110 h at −183 to −192°C *(6)*. In that study, air-dried seeds were used at a natural moisture level of 10–12%. In the intervening period it has been demonstrated that dry seeds of about 300 species tolerate exposure to liquid nitrogen temperature *(2,3, 7–10a)*, including representatives of some important plant families, such as the Palmae *(11)* and Orchidaceae *(4,12)*. Moreover, it is expected that a majority of orthodox seeds of other species will respond favorably to cryopreservative treatment. However, there are plant species for which seed storage, under conventional or cryopreservation conditions, is not feasible. For example, for species that reproduce clonally or do not readily produce viable seed, meristem cryopreservation is an alternative strategy for their *ex situ* preservation. Also, many woody perennials produce large fleshy seeds that are intolerant of drying to low seed moisture contents (<20%), for example, species of oak and dipterocarp. For these "recalcitrant" seeded species, techniques are being developed for the cryopreservation of the embryo, including the excised embryonic axis. This approach is beyond the scope of this chapter, but introductions to the topic are available *(13,14)*. An alternative approach to the problem of recalcitrant seed cryopreservation has been to simulate their response to cryoprotection and freezing using rehydrated orthodox seeds *(15,16)*. However, the validity of this approach is uncertain; a recalcitrant seed is unlikely to have undergone the same physiological and biochemical modifications associated with the early stages of orthodox seed maturation drying. Nonetheless, this approach to the cryopreservation of orthodox seed is addressed here. In addition, this chapter summarizes the main factors that need to be controlled to achieve the successful cryopreservation of dry seeds: seed moisture content and the cooling/warming regimes used. Finally, attention will be given to the methods for assessment of seed quality following cryopreservation. In this chapter cryopreservation is defined as "the preservation or storage in very cold temperatures, usually in liquid nitrogen" *(17)*, at temperatures close to −196°C.

2. Materials
2.1. Freezing of Dry Seeds

1. Seeds: Seeds should be dried to low moisture content by exposure to low relative humidity air. Large-scale drying rooms, often running at around 15–20% relative humidity, can be used to dehydrate bulk quantities of seed. Smaller seedlots can be dried over saturated salt solutions within hermetically sealed containers; plastic sandwich boxes or Kilner jars are ideal. To prepare the salt solution, add excess solid to a small quantity of water. The ratio will vary with each particular salt. Boil the solution to ensure maximum solubility and transfer to equilibration/drying chamber. To be effective, the saturated solution should always contain a large quantity of undissolved salt. Examples of the salts used to generate particular relative humidities are given in Table 1, with an indication of how seed oil content influences the moisture contents to which the seed would equilibrate (*see* Note 1).

Table 1
Constant Humidity Solutions and Predicted Seed Equilibrium Moisture Contents[a]

Salt	Relative humidity, percentage[b]	Seed moisture content, percentage[c]	
		Pea	Sesame
$LiCl \cdot 2H_2O$	11	4.9	2.7
$LiI \cdot 3H_2O$	18	6.3	3.4
$CaCl_2 \cdot 6H_2O$	29	8.1	4.5
$NaI \cdot 2H_2O$	38	9.4	5.2
$Mg(NO_3)_2 \cdot 6H_2O$	53	11.5	6.5
KI	69	14.0	8.0
NaCl	75	15.0	8.6
$(NH_4)2SO_4$	81	16.2	9.3
$BaCl_2 \cdot 2H_2O$	90	18.5	10.8

[a]Fresh weight basis at 25°C for pea (1.5% oil, dry wt basis) and sesame (47.5% oil).
[b]From *(18)*.
[c]Calculated from *(19)* assuming seed desorption. It was suggested that the mathematical relationship between seed moisture content, oil content, temperature, and relative humidity was applicable up to a humidity of 70% *(19)*. Thus, the moisture content values given above for higher humidities must be considered provisional, although comparison to published sorption isotherm data suggest the error of prediction is only about 10%.

2. Stainless steel or aluminum weighing dishes with lids.
3. Fan oven.
4. A dry room or relative humidity chamber.
5. Cryostorage containers: Screwcap polypropylene ampules with gasket, borosilicate vials with neoprene insert and crimped aluminum caps, aluminum foil laminate bags (with double-sealed edges), and polyolifin tubing are suitable. The most appropriate container to use will depend on the volume of the seed for storage. Endeavor to limit the container's cross-sectional area to ensure relatively uniform cooling and warming rates throughout it.
6. A programmable freezer (*see* Note 2).
7. Storage system. Small seed samples are easily accommodated in drawer-based inventory systems; larger seed samples suit storage in canisters (*see* Note 3).
8. A water bath preset to ≈40–45°C (*see* Note 4).
9. 1% (w/v) Triphenyl tetrazolium chloride (TTC) in phosphate buffer is required for the vital, histochemical staining test (*see* Note 5):
    a. Dissolve 3.631 g KH_2PO_4 in 400 mL of distilled water;
    b. Dissolve 7.126 g $Na_2HPO_4 \cdot 2H_2O$ in 600 mL of distilled water;
    c. Mix the two solutions; and
    d. Dissolve 10 g of 2,3,5-TTC in 1.0 L of the buffer solution (*see* Note 5).

10. Dissecting instruments.
11. Incubator with temperature and lighting control.
12. Protective equipment: cryogloves, goggles, and so on.

2.2. Freezing of Cryoprotected Seeds

1. Seed rehydration: Moistened filter paper, paper towels, or 1% (w/v) agar-water in Petri dishes (*see* Note 6).
2. Cryoprotectants are only required for seeds that have high water content. The cryoprotectant is usually applied singly after seed rehydration has taken place, possible chemicals include dimethyl sulfoxide (DMSO), glycerol, or 1,2-propanediol (*see* Notes 7–9).
3. Commercially available sodium hypochlorite solution (*see* Note 10). Other materials as detailed in Section 2.1.

3. Methods
3.1. Freezing of Dry Seeds

1. Determine moisture content of cleaned seed sample by the following protocol: Weigh seeds before and after drying in a fan-driven oven (17 h at 103°C for oily seeds; 2 h at 130°C for nonoily seeds). Express moisture content on a fresh weight basis.
2. Determine initial viability of seeds using the appropriate germination test conditions or by tetrazolium staining. The tetrazolium test procedure includes the following basic steps:
   a. Rehydrate seed over water at 25°C for a few days;
   b. Penetrate covering seed structures or surgically remove embryo;
   c. Soak seed/embryo individually in a small volume ($\approx$2 mL) of TTC for 24–48 h in the dark at 30°C;
   d. Wash seed/embryo several times in distilled water; and
   e. Evaluate immediately after washing or following brief storage of stained material on moistened filter paper at 3–5°C. Note the intensity and topography of the red staining pattern on the embryo tissues, i.e., embryonic axis and cotyledon. No stain indicates no viability (*see* Note 11).
3. Place seed in the dry room or relative humidity cabinet as a monolayer (for uniform drying of all seeds). Preweigh a small sublot of seed and include with the main seedlot as a moisture content monitor. Estimate the dry weight of the monitor seed from the original moisture content of the sample. Weigh seeds regularly during drying and produce an estimate of seed moisture by deducting the estimated dry weight from each fresh weight determination and dividing the resultant weight of water by the seed fresh weight at each sample time during desiccation. Drying should proceed to below the high moisture freezing limit (HMFL) of the seeds before their transfer to liquid nitrogen can be considered (*see* Note 12).

4. Remove seed from the drying environment when moisture content has been reduced to the desired level (*see* Note 13).
 5. Reassess germination level/stainability of seed. If test is negative seed may be sensitive to drying, i.e., is recalcitrant.
 6. Either count number of dried seed for storage or take total weight of sample and 100 seed weight and calculate total number of seed in sample.
 7. Place small quantities of dry seeds (i.e., grams) and seal in appropriate containers then cool at about 10°C/min (*see* Note 14).
 8. Transfer to cryostorage vessel.
 9. To thaw: Rewarm containers in a water bath at approx 40°C if seeds are close to their HMFL or at ambient laboratory conditions if seed is drier.
 10. Allow seed to equilibrate to room temperature before assessing viability.
 11. Perform viability test: Two approaches can be followed, vital histochemical staining (*see* step 2) or seed germination (*see* Note 15). For germination, use moistened filter paper or, preferably, 1% (w/v) agar gelled in distilled water. Do not rapidly rehydrate dry seeds (i.e., <30% relative humidity) by soaking in water as seeds of some species are sensitive to imbibitional injury (*see* Notes 16 and 17).

3.2. Freezing of Cryoprotected Seed

 1. Rehydrate seed on moist filter paper, or agar, or above water to desired moisture content at room temperature (*see* Note 18).
 2. Surface sterilize the seed in a dilute solution of commercial sodium hypochlorite (about 1% available chlorine) for 30 min. Higher concentrations and longer times may be necessary depending on the level of coat-borne infections and surface topography.
 3. Following rehydration, soak the seed in cryoprotectant for 1 h at 25 or 4°C (*see* Note 19).
 4. Remove seed from cryoprotectant, surface dry on filter/tissue paper. Transfer to appropriate container and place directly in liquid nitrogen, then transfer to cryostorage vessel.
 5. Thaw in water bath at 40–45°C for 30 s–2 min or in air at room temperature for 20 min as appropriate.
 6. Sow seeds for germination without removal of cryoprotectant, although when high concentrations have been employed, remove by washing. Alternatively assess seed quality by tetrazolium staining.

4. Notes

 1. After drying, the moisture content of the seedlot should be confirmed gravimetrically before freezing. Fan-assisted ovens are required to drive off the residual seed water. Operational temperatures of 103 and 130°C are suitable for oily and non-oily seed, respectively.

2. Alternatively, similar cooling-rates can be achieved through the use of a differing thicknesses of insulation materials (e.g., ampules inside layers of aluminum) and by altering the distance of the sample from the surface of the liquid nitrogen. It is possible to improvise with various wide-neck dewars of varying depth. Always chart the cooling (and rewarming) regimes using a copper-constantan, or similar, thermocouple and electronic thermometer.
3. Storage in the vapor phase is more convenient if seed samples are to be removed on a frequent basis. Storage in the liquid phase does, however, reduce the possibility of inadvertent warming, but extreme care must be exercised when removing samples from the liquid; defective containers that allow the penetration of liquid may explode during rewarming as the liquid expands to become a gas within a confined space. Always wear appropriate protective clothing (goggles, gloves, apron) and use long metal forceps for the transfer of samples between dewars.
4. No particular apparatus is required for the rewarming of the dry seed samples as it is usual to rewarm under laboratory conditions. However, if the frozen seed has a moisture content close to its high moisture freezing limit seeds should be thawed in a heated water bath.
5. TTC is a mild irritant and gloves should be worn during its preparation, which should take place under a fume hood. Store the prepared solution in the dark as it is light sensitive; UV light stimulates a slow reduction of the salt. Also store cool, but allow to warm up to room temperature before use to avoid any likelihood of chilling injury in tropical seeds especially. The prepared solution should remain effective over a few weeks. Dissecting instruments are required to facilitate the penetration of the stain into the seed. Use Petri dishes with individual wells. Wrap the dishes in aluminum foil as the development of the staining pattern must take place under dark conditions.
6. Although soaking in water has often been used as the method to raise seed moisture content before cryoprotectant addition *(15,16)*, it is not recommended as a general seed rehydration treatment.
7. Alternatively, cryoprotectants mixtures can be used in the same way, for example, combinations of 1,2-propanediol, DMSO, and sucrose *(16)*. Another combination of protectants, consisting of glucose, polyethylene glycol 6000, and DMSO, has been used as the rehydration medium for seeds before cryopreservation, although to no apparent benefit *(11)*.
8. Always use Analar-grade reagents, except DMSO, which should be spectroscopy grade. Solutions are made up in distilled water and should be used soon after preparation, although short-term storage in the refrigerator is possible.

Table 2
Seed HMFL and Seed Oil Contents of Selected Species

Species	Oil content, percentage dry wt[a]	HMFL percentage fresh weight[b]	Refs.
Sesame	50–55	9	3
Radish	≈40	17	3
Lettuce	34–42	18–20	5
Cucumber	32	16	3
Cabbage	26	14	3
Tomato	18–27	19	3
Onion	20	25	3
Soybean	18	14	7
Carrot	13	22	3
Alfalfa	12	17	7
Barley	2	21	3
Italian ryegrass	2	21	7
Wheat	2	27	3

[a]Values taken from refs. *(23,24)*.
[b]Signifies the moisture content above which the seed viability would be expected to fall significantly on transfer to liquid nitrogen. The values are dependent to a certain extent on the cooling/warming regime used, and this has varied between studies; generally, rapid cooling and rewarming has been used.

9. DMSO freely penetrates the skin, can be irritating to the skin and eyes, and may be cytotoxic (concentration and exposure time-dependent). Always wear protective clothing, including nitrile gloves. Information on the correct precautions to take when handling chemicals can be found in compendium volumes containing "hazard data sheets," such as the one produced by BDH Limited (Poole, UK).
10. Wet seeds are likely to possess an enhanced surface microflora and thus these seeds are best surface sterilized before freezing using a solution of commercially available sodium hypochlorite.
11. An indication of the appropriate germination and tetrazolium staining conditions for seeds of many different species is contained in refs. *(20,21)*. When dealing with the cryopreservation of new species some empirical studies may be required to develop optimum viability test conditions. The assessment of tetrazolium staining can be subjective and the determination of the correlation between this test, which reflects dehydrogenase enzyme activity, and the germination test, is recommended for each species.
12. The HMFL varies with species (Table 2) and is inversely related to seed oil content (HMFL = 23.1 + (–0.21 oil); significant at $p = 0.01$). Based on differential scanning calorimetry studies on the properties of water in pea

Table 3
Optimum Equilibrium Oilseed Water Status For Survival of Rapid Freezing[a]

Species	Oil content, percentage dry wt[b]	Moisture content, percentage fresh weight	Relative humidity, percentage[c]
Soybean	18	10.4–10.6[d]	58–59
Sunflower	33	8.3–10.0[e]	53–69
Niger/Noog	35	10.7[f]	77
Groundnut	45	6.7–7.4[d]	53–60
Sesame	≈48	8.0–10.0[g]	70–85

[a]At 25°C for subsequent safe exposure to liquid nitrogen temperatures at a cooling rate of ≈200°C/min. Information sources: [b]From (23). [c]Calculated from (19). [d]From (25). [e]From (26). [f]From (27). [g]From (3).

and soybean seeds (22), this upper moisture limit for cryopreservation probably coincides with a relative humidity range of 80–92%.

13. Optimum moisture contents for the cryopreservation of oilseeds using rapid cooling (≈200°C/min) are given in Table 3; optima are less evident in non-oily seeds treated in the same way. Although the optima vary between species, this is largely owing to the variation in oil content as the optimal relative humidity to which seeds should be dried appears quite similar between species; 66 ± 5% on the average. Recently, it has been theorized on thermodynamic grounds that the optimum moisture content for seed storage at –150°C is equivalent to relative humidities at room temperature of between 61 and 72% (27a).

At moisture contents below this optimum, some seeds can exhibit physical injury following rapid cooling/warming. The injury may manifest in many ways: fracturing of the embryo tissue (7), including the detachment of the embryo cotyledons (3,28); production of seedlings with abnormalities, such as adventitious root growth in sesame (29); and split radicles in maize (30). In lipid-rich seeds of sunflower and soybean, it has been proposed that reduced seed vigor, i.e., germination rate, is caused when seed lipids are induced to undergo a glass transition during rapid cooling; it is possible that the subsequent cracking of the glasses causes mechanical injury (26,31).

At moisture contents above this optimum, the moisture status of the seeds will be close to the HMFL and ice formation may be a problem during cooling or rewarming. Small quantities of ice in the seed may be tolerated, but are more likely to become damaging in lipid-rich seeds as lipids undergoing thermal transitions may enable the coalescence of small, benign ice crystals into larger pernicious ones (31).

14. Physical injury to dry seeds (e.g., equilibrated to 15% relative humidity) can occur at cooling rates as slow as 20°C/min *(28)*. However, cooling rates around 10°C/min overcome this problem *(3,26)*.
15. Measurement of germination level is a relatively slow procedure (days to weeks) if seed dormancy is present in the seedlot, but is far more reliable than vital staining. The staining test does, however, have the benefit of being more rapid (about 2 d to complete). When used together the two tests are complementary, as the vital staining technique may yield information on the viability of the seedlot before any emergence has occurred in the germination test.
16. Slow rehydration can be achieved using high relative humidity, for example at 90% using a saturated salt solution (Table 1) or at 100%, above water. Environmental requirements for germination will vary with species and a range of controlled temperature incubators with alternating temperature and light cycling features are required. At the end of the germination test, a qualitative assessment of seed viability may be necessary if no germination has occurred. Dissecting instruments (scalpel and fine forceps) are required to distinguish soft, moldy, inviable seeds from firm, viable seeds.
17. Seeds of some species may increase their germination rate or level after cryopreservation. In the case of *Trifolium arvense* seeds, an increase in germination rate after treatment was ascribed to the removal of hard seededness, i.e., seed impermeability to water was lowered *(28)*. In contrast, physical damage to the seed coat was not thought to be implicated in the increase of *Setaria lutescens* seed germination after freezing; dissociation of the embryo lipids was one proposed cause of this effect *(32)*.
18. Seed rehydrated to 24–42% moisture content has been demonstrated to survive cryopreservation after cryoprotection *(15,16)*. This step is usually performed around room temperature *(10,15,16)* or may be performed at 4°C *(16)*. Time of rehydration will vary with the permeability characteristics of the seed; determine empirically for each species.
19. Most success has been achieved using DMSO at 15% (v/v) *(10,15)* and 35% (v/v) *(10)*, and 1,2-propanediol combined with sucrose (38 and 20% [v/v], respectively) *(16)*.

References

1. Dickie, J. B., Ellis, R. H., Kraak, H. L., Ryder, K., and Tompsett, P. B. (1990) Temperature and seed storage longevity. *Ann. Bot.* **65**, 197–204.
2. Stanwood, P. C. and Bass, L. N. (1981) Seed germplasm preservation using liquid nitrogen. *Seed Sci. Technol.* **9**, 423–437.
3. Stanwood, P. C. (1985) Cryopreservation of seed germplasm for genetic conservation, in *Cryopreservation of Plant Cells and Organs* (Kartha, K. K., ed.), CRC, Boca Raton, FL, pp. 199–226.

4. Pritchard, H. W. and Seaton, P. T. (1993) Orchid seed storage: historical perspective, current status, and future prospects for long-term conservation. *Selbyana* **14**, 89–104.
5. Roos, E. E. and Stanwood, P. C. (1981) Effects of low temperature, cooling rate, and moisture content on seed germination of lettuce. *J. Am. Soc. Hort. Sci.* **106**, 30–34.
5a. Iriondo, J. M., Pérez, C., and Pérez-García, F. (1992) Effect of seed storage in liquid nitrogen on germination of several crop and wild species. *Seed Sci. Technol.* **20**, 165–171.
6. Brown, H. T. and Escombe, F. (1897–1898) Note on the influence of very low temperatures on the germinative power of seeds. *Proc. Royal Soc. Lond.* **62**, 160–165.
7. Sakai, A. and Noshiro, M. (1975) Some factors contributing to the survival of crop seeds cooled to the temperature of liquid nitrogen, in *Crop Genetic Resources for Today and Tomorrow* (Frankel, O. H. and Hawkes, J. C., eds.), Cambridge University Press, Cambridge, UK, pp. 317–326.
8. Stanwood, P. C. and Roos, E. E. (1979) Seed storage of several horticultural species in liquid nitrogen (–196°C). *Hort. Sci.* **14**, 628–630.
9. Styles, E. D., Burgess, J. M., Mason, C., and Huber, B. M. (1982) Storage of seed in liquid nitrogen. *Cryobiology* **19**, 195–199.
10. Pence, V. C. (1991) Cryopreservation of seeds of Ohio native plants and related species. *Seed Sci. Technol.* **19**, 235–251.
10a. Touchell, D. H. and Dixon, K. W. (1993) Cryopreservation of seed of Western Australian native species. *Biodiversity Conserv.* **2**, 594–602.
11. Al-Madeni, M. A. and Tisserat, B. (1986) Survival of palm seeds under cryogenic conditions. *Seed Sci. Technol.* **14**, 79–85.
12. Pritchard, H. W. (1984) Liquid nitrogen preservation of terrestrial and epiphytic orchid seed. *Cryo-Lett.* **5**, 295–300.
13. Pritchard, H. W. and Prendergast, F. G. (1986) Effects of desiccation and cryopreservation on the in vitro viability of embryos of the recalcitrant seed species *Araucaria hunsteinii* K. Schum. *J. Exp. Bot.* **37**, 1388–1397.
14. Pence, V. C. (1992) Desiccation and the survival of *Aesculus, Castanea,* and *Quercus* embryo axes through cryopreservation. *Cryobiology* **29**, 391–399.
15. Grout, B. W. W. (1979) Low temperature storage of imbibed tomato seeds: a model for recalcitrant seed storage. *Cryo-Lett.* **1**, 71–76.
16. de Boucaud, M.-T. and Cambecedes, J. (1988) The use of 1,2-propanediol for cryopreservation of recalcitrant seeds: the model case of *Zea mays* imbibed seeds. *Cryo-Lett.* **9**, 94–101.
17. International Board for Plant Genetic Resources (1991) *Elsevier's Dictionary of Plant Genetic Resources.* Elsevier, Amsterdam.
18. Wexler, A. (1993–1994) Constant humidity solutions, in *CRC Handbook of Chemistry and Physics,* 74th ed. (Lide, D. R., ed.-in-chief), CRC, Boca Raton, FL, pp. 15–25.
19. Cromarty, A. S., Ellis, R. H., and Roberts, E. H. (1985) *Design of Seed Storage Facilities for Genetic Conservation.* International Board for Plant Genetic Resources, Rome.

20. Ellis, R. H., Hong, T. D., and Roberts, E. H. (1985) *Handbook of Seed Technology for Genebanks. Volume I. Principles and Methodology.* International Board for Plant Genetic Resources, Rome.
21. Ellis, R. H., Hong, T. D., and Roberts, E. H. (1985) *Handbook of Seed Technology for Genebanks. Volume II. Compendium of Specific Germination Information and Test Recommendations.* International Board for Plant Genetic Resources, Rome.
22. Vertucci, C. W. (1990) Calorimetric studies of the state of water in seed tissues. *Biophys. J.* **58,** 1463–1471.
23. Eckey, E. W. (1954) *Vegetable Fats and Oils.* Reinhold Publishing Corporation, New York.
24. Earle, F. R. and Jones, Q. (1962) Analyses of seed samples from 113 plant families. *Econ. Bot.* **16,** 221–250.
25. Zhang, B., Fu, J.-R., and Xu, S. (1990) Studies on cryopreservation of seeds of crops and vegetables. *Acta Scientiarum Naturalium Universitatis Sunyatseni* **29,** 115–121.
26. Vertucci, C. W. (1989) Effects of cooling rate on seeds exposed to liquid nitrogen temperatures. *Plant Physiol.* **90,** 1478–1485.
27. Zewdie, M. and Ellis, R. H. (1991) Survival of tef and niger seeds following exposure to sub-zero temperatures at various moisture contents. *Seed Sci. Technol.* **19,** 309–317.
27a. Vertucci, C. W. and Roos, E. E. (1993) Theoretical basis of protocols for seed storage II. The influence of temperature on optimal moisture levels. *Seed Sci. Res.* **3,** 201–213.
28. Pritchard, H. W., Manger, K. R., and Prendergast, F. G. (1988) Changes in *Trifolium arvense* seed quality following alternating temperature treatment using liquid nitrogen. *Ann. Bot.* **62,** 1–11.
29. Stanwood, P. C. (1980) Tolerance of crop seeds to cooling and storage in liquid nitrogen (–196°C). *J. Seed Technol.* **5,** 26–31.
30. Harrison, B. J. and Carpenter, R. (1977) Storage of *Allium cepa* seed at low temperatures. *Seed Sci. Technol.* **5,** 699–702.
31. Vertucci, C. W. (1989) Relationship between thermal transitions and freezing injury in pea and soybean seeds. *Plant Physiol.* **90,** 1121–1128.
32. Jordan, J. L., Jordan, L. S., and Jordan, C. M. (1982) Effects of freezing to –196°C and thawing on *Setaria lutescens* seeds. *Cryobiology* **19,** 435–442.

CHAPTER 15

Cryopreservation of the Sperm of the Pacific Oyster *Crassostrea gigas*

Iain R. B. McFadzen

1. Introduction

In contrast to the extensive research and advances in the cryopreservation of animal cell lines and tissues, few methodologies are available for the cryopreservation of invertebrate cells. Several research groups have reported methodologies for marine invertebrate gametes *(1–3);* these predominantly concentrate on the spermatozoa of the Pacific oyster *(Crassostrea gigas).* Marine invertebrate gametes, embryos, and larvae have been extensively deployed to monitor marine pollution and to assess the biological quality of coastal and oceanic waters *(4–7).* This interest stems from the relative ease of their laboratory culture and the sensitivity of the early developmental stages to environmental pollutants, when deployed in biological effects monitoring programs *(4,5,7).* The use of cryopreserved oyster larvae as material for use in bioassays of water quality has recently been advocated *(7).*

Additionally the cryopreservation of the spermatozoa of the Pacific oyster (a protandric hermaphrodite) has enabled researchers to maintain a ready supply of gametes for successful self-fertilization studies *(3).* Guaranteed supplies of sperm, without seasonal limitations and costly hatchery maintenance of adult broodstocks, increase the possibilities for selective inbreeding programs. Selection for desirable traits, such as disease resistance and increased growth potential, are made possible with the additional opportunity to maintain stocks that are suitable for ploidy manipulations, i.e., triploids and tetraploids (potential benefits to commercial farming).

A simple, effective protocol is presented in this chapter, which combines the use of adult conditioning, cryoprotective agents, controlled cooling rates and a thawing regime, to maximize postthaw viability, enabling researchers to cryopreserve Pacific oyster spermatozoa. This technique limits the damage induced by changes in the composition of intra- and extracellular solutions, and external stresses incurred as a result of ice formation, whereas emphasis is placed on the necessity to produce high quality broodstock prior to freezing.

2. Materials

1. Cryoprotective agent (CPA) stock solution (*see* Notes 1 and 2): $1M$ trehalose (dihydrate form) dissolved in cool distilled water, 10% (v/v) dimethyl sulfoxide (DMSO) (*see* Note 3). Oxygen saturate the cryoprotectant by passing compressed air through the solution for 15 min prior to use.
2. A controlled rate freezer with accompanying pressurized 30 L liquid nitrogen dewar.
3. Storage dewar with 30 L liquid nitrogen.
4. 5-L Tall dewar for initial plunge.
5. Metal rack for 0.5-mL plastic cryostraws (Instruments de Medicine Veterinaire [IMV], l'Aigle, France).
6. Water bath.
7. Safety equipment: long forceps, cryogloves, cryoapron, protective full-face screen.
8. Source of gametes: Gametes are obtained from gravid adult broodstock (*see* Section 3., step 2), which have been conditioned in the laboratory *(8)*.

3. Method

1. Conditioning of adult oysters: Adult oysters require a period of conditioning in the laboratory or hatchery in order to complete gametogenesis, particularly in temperate regions where seasonal variations in ambient water temperature exist. Key aspects of broodstock conditioning *(8)* are typically high food rations (mixed algal diet) and warm water conditions (*see* Note 4).
2. Stripping of adult oysters: Extract gametes from individual adult oysters after carefully removing the upper valve of the oyster shell (*see* Note 5). Gently remove the gametes from the gonad using a clean Pasteur pipet. Make a small incision in the gonad and insert the tip of the pipet directly beneath the gonad membrane while maintaining positive pressure on the pipet bulb. Once the tip of the pipet is located within the gonad, gradually release the pressure on the bulb and ease the gametes into the stem of the

pipet (*see* Note 6), repeat the procedure until the desired volume of concentrated spermatozoa has been collected.
3. CPA incubation of the spermatozoa: Concentrate spermatozoa from individual adults in a clean glass pipet directly from the gonadal tissue and place immediately into the cryoprotective agent at ambient temperature; incubate for a total of 15 min (*see* Notes 7 and 8). Mix equal volumes of spermatozoa and CPA to give a sufficient volume of material to fill approx 10 straws (5 mL total volume) from each adult (*see* Note 9).
4. Loading of the cryostraws: Thoroughly mix the spermatozoa and cryoprotectant mixture by gentle agitation. Then rapidly draw the solution into the 0.5–mL plastic straws (IMV) (*see* Note 10). Prepare a minimum of 10 straws for each cooling run; a maximum of 50 straws are recommended for this particular protocol and controlled rate freezer (*see* Note 11).
5. Arrange the straws on precooled (4°C) trays so they sit perpendicular to the vapor flow within the cooling chamber.
6. Cooling: Using a controlled rate freezer use the following cooling regime (*see* Note 12).
   a. Ramp 1: From 25°C, cool the chamber at 100°C/min to a chamber temperature of –120°C.
   b. Ramp 2: From –120°C, cool the chamber at 15°C/min to a chamber temperature of –150°C.
   c. Ramp 3: Maintain a steady chamber temperature of –150°C for 1 min.
7. Plunge the straws into sufficient liquid nitrogen to cover them (*see* Note 13).
8. Store the straws under liquid nitrogen (–196°C).
9. To recover the spermatozoa: Carefully remove individual frozen straws from the storage medium (liquid nitrogen) using long forceps and plunge directly into a water bath no warmer than the desired rearing temperature, 25°C maximum (*see* Note 14). Pour the contents of the straws into an equal volume of UV-treated seawater (32–34 PSU), then agitate for 2 min prior to assessing postthaw motility (*see* Note 15).

4. Notes

1. Reagents should be prepared no more than 2 h before use and stored at the ambient hatchery temperature, normally at 25°C. All glassware and filtration equipment should be clean and free from bacteria, as any bacterial contamination during the incubation period will be detrimental to subsequent postthaw survival.
2. The addition of DMSO to the trehalose solution is an exothermic reaction, which if not controlled prior to addition of the spermatozoa, will result in denaturing the sperm head. Following the addition of DMSO to the trehalose, the solution should be vigorously aerated to obtain oxygen satura-

tion. During this stage the CPA should be allowed to cool to the desired temperature (25°C) before use.
3. DMSO is cytotoxic and care should be taken when handling the CPA.
4. Adult conditioning: If incoming seawater quality can not be guaranteed (possibly a seasonal variable), then provision should be made to filter (0.45–2.0 µm) or irradiate with UV light. In order to successfully cryopreserve viable spermatozoa, the adult oyster must be in a mature gravid condition prior to removal of the gametes and the initiation of the subsequent freezing protocol.
5. The outer surface of all adult oysters should be scrubbed clean and rinsed in fresh seawater before gametes are removed. This will reduce the risk of infection from ciliates and other protozoa often located on the adult shell.
6. Prime the pipet tip with a small volume of CPA before attempting to remove the gametes. This will prevent the thick, viscous sperm from adhering to the glass.
7. If simultaneously cryopreserving gametes from more than one adult, ensure a reliable form of straw identification is employed. The identifying marks or code should be readily distinguishable through liquid nitrogen and associated vapor.
8. Incubation of spermatozoa in CPA: Loading of straws takes approx 2–3 min, depending on the total number required. Allowance should be made for loading of the straws during the incubation period, and the total precooling incubation time of 15 min should not be exceeded.
9. The volume of spermatozoa produced by individual oysters is highly variable. Generally no more than 50% of the gonad volume is extracted for freezing, this reduces the chance of removing immature or nutritive cells and of piercing the digestive gland (located in the middle of the gonad) with the pipet tip.
10. Wipe the wet end of the straws with tissue to prevent them freezing together during the cooling phase.
11. The cooling rate described is based on a maximum of 50 cryostraws frozen in a single batch using a Planer Kryo 10 Mk II controlled rate freezer. Any differences in chamber size or total volume of material being frozen, may result in variations in the desired cooling profile.
12. Preparation of the cooling chamber: A minimum of three trial cooling runs should be conducted to ensure that the connecting pipes and surrounding insulation to the chamber are sufficiently cooled prior to initiating a true run. Failure to precool the chamber can result in the surrounding insulation materials acting as a heatsink, with the result of a false cooling profile being achieved.

13. Plunging into liquid nitrogen: This step should be carried out as quickly as possible with minimum handling of the straws. Removal of the chamber lid must be swift and only once the desired chamber temperature has been maintained. Transfer of straws from the chamber racks to the dewar must be rapid and ample liquid nitrogen should be in the dewar to ensure the straws will be covered. Allowance should be made for the loss in volume as a result of liquid nitrogen boiling off on straw immersion.
14. Thawing: The water bath should not exceed 25°C as a rise in temperature above the holding temperature of the initial broodstocks will denature the spermatozoa. Thawed straws should then be removed from the bath as soon as the ice has melted. If using freshwater to thaw the straws, ensure that the outside of the straws are dried with tissue to remove excess water, thus avoiding osmotic stress to the sample.
15. Overall postthaw motility may be reduced compared to fresh spermatozoa, however, viability should be assessed by fertilization success after addition to fresh oyster eggs. Polar body formation or the appearance of the first cleavage furrow in the developing embryo can be regarded as an indicator of a successful fertilization.

References

1. Zell, S. R., Bamford, M. H., and Hidu, H. (1979) Cryopreservation of spermatozoa of the American oyster *Crassostrea virginica*. *Cryobiology* **16**, 448–460.
2. Bourgrier, S. and Rabenomanana, L. D. (1986) Cryopreservation of the spermatozoa of the Japanese oyster, *Crassostrea gigas*. *Aquaculture* **58**, 277–280.
3. Yankson, K. and Moyse, J. (1991) Cryopreservation of the spermatozoa of *Crassostrea tulipa* and three other oysters. *Aquaculture* **97**, 259–267.
4. Woelke, C. E. (1972) Development of a receiving water quality bioassay criterion based on the 48 hour Pacific oyster *(Crassostrea gigas)* embryo. Washington Department of Fisheries, Technical Report No. 9, 1–93.
5. Stebbing, A. R. D. (1985) Bioassay, in *The Effects of Stress and Pollution on Marine Animals* (Bayne, B. L., Brown, D. A., Burns, K., Dixon, D., Ivanovici, A., Livingstone, R., Lowe, D. M., Moore, M. N., Stebbing, A. R. D., and Widdows, J., eds.), Praeger, New York, pp. 133–137.
6. Thain, J. E. (1991) *Biological Effects of Contaminants: Oyster (Crassostrea gigas) Embryo Bioassay.* International Council for the Exploration of the Sea report. Techniques in Marine Environmental Sciences No. 11, Copenhagen.
7. McFadzen, I. R. B. (1993) Growth and survival of cryopreserved oyster and clam larvae along a pollution gradient in the German Bight. *Mar. Ecol. Prog. Ser.* **91**, 215–220.
8. Utting, S. D. and Spencer, B. E. (1991) *The Hatchery Culture of Bivalve Mollusc Larvae and Juveniles.* Ministry of Agriculture Fisheries and Food. Directorate of Fisheries Research, Lowestoft Laboratory Leaflet No. 68.

CHAPTER 16

Cryopreservation of Fish Spermatozoa

Krishen J. Rana

1. Introduction

Cryopreservation of fish spermatozoa has closely shadowed historical developments of the preservation technology used for mammalian gametes. Within 4 yr of the discovery of the cryoprotective value of glycerol *(1)*, slices of Atlantic herring *(Clupea harengus)* testis were successfully cryopreserved and stored in dry ice for 6 mo in 80% (v/v) seawater containing 12.5% (v/v) glycerol *(2)*.

Since then, and notably in the 1970s, there have been a flurry of publications demonstrating the feasibility of cryopreserving fish spermatozoa from a large number of marine and freshwater species. To date, spermatozoa of over 50 species of freshwater and marine fish have been cryopreserved, and many of these studies have been reviewed in several recent publications *(3–8)*. A number of different protocols are advocated in the literature for the preservation fish spermatozoa. The majority of publications and protocols, however, relate to three groups of fish of aquacultural importance: the salmonoids, tilapia, and carp.

The sophistication of techniques and equipment used in fish spermatozoa cryopreservation parallels that of cattle, with some notable exceptions. Extenders are required for diluting fish milt prior to cryopreservation. Such solutions, which are generally designed to be compatible with the physicochemical composition of the seminal plasma of the candidate species, are tailored to maintain the spermatozoa in an immotile but viable state, until required *(3,4)*. To ameliorate cryoinjuries both permeating and nonpermeating cryoprotectants have been advocated. Although glycerol, dimethyl sulfoxide (DMSO), and methanol (MeOH) have been used, DMSO and lately MeOH are generally the cryoprotectants of choice. To minimize cryoinjuries, nonpermeating cryoprotectants, such as skimmed milk powder, egg yolk, proteins, and sugars, have also been included in extender receipts. The chemical constituents of cryodiluents used for the cryopreservation of fish spermatozoa varies enormously *(3,6)*. The value of complex cryodiluents for fish cryopreservation remains unclear since equally good results have been reported with simple mixtures of sugars and DMSO *(9,10)*. The vast majority of simpler extenders are composed of variable combinations of chloride salts of sodium, potassium magnesium, and calcium and sodium or potassium bicarbonate.

Various cooling methods and apparatus have been successfully applied for cryopreserving fish spermatozoa. Milt is commonly packaged in cryovials *(11–14)* or plastic straws *(15–17)* cooled over liquid nitrogen vapor column and stored in liquid nitrogen. For field applications and rapid freezing, milt can be cryopreserved as pellets on dry-ice blocks and then stored in capped cryovials *(9,10,18,19)*. The cooling rates of milt generated between samples and studies using these methods, however, are often unknown, and if known they tend to be variable and unpredictable (Fig. 1). The advent of controlled rate coolers (CRCs) and plastic straws for rapid heat transfer facilitated the reproducibility of cooling regimes during cryopreservation studies and provided the means for establishing optimal cooling rates for fish spermatozoa. With such systems, the overall success of postthaw viability of spermatozoa for species, such as the tilapias are consistently similar to those obtained with fresh milt *(14)*.

The interpretation of results for fish spermatozoa cryopreservation are fraught with difficulties. Despite the large number of publications over the last 25 yr, the various components of cryopreservation technology, such as prefreezing milt quality, packaging, cooling, thawing, insemination, and evaluation of techniques used between and within studies, vary considerably. Consequently, the results for similar protocols and species vary widely and their practical value in evaluating and standardizing cryopreservation technology for genebanking has been greatly compromised.

Much of the earlier ambiguity in the literature stemmed from the direct transfer of protocols from mammalian spermatozoa preservation to fish without due consideration to the differences between their sperm

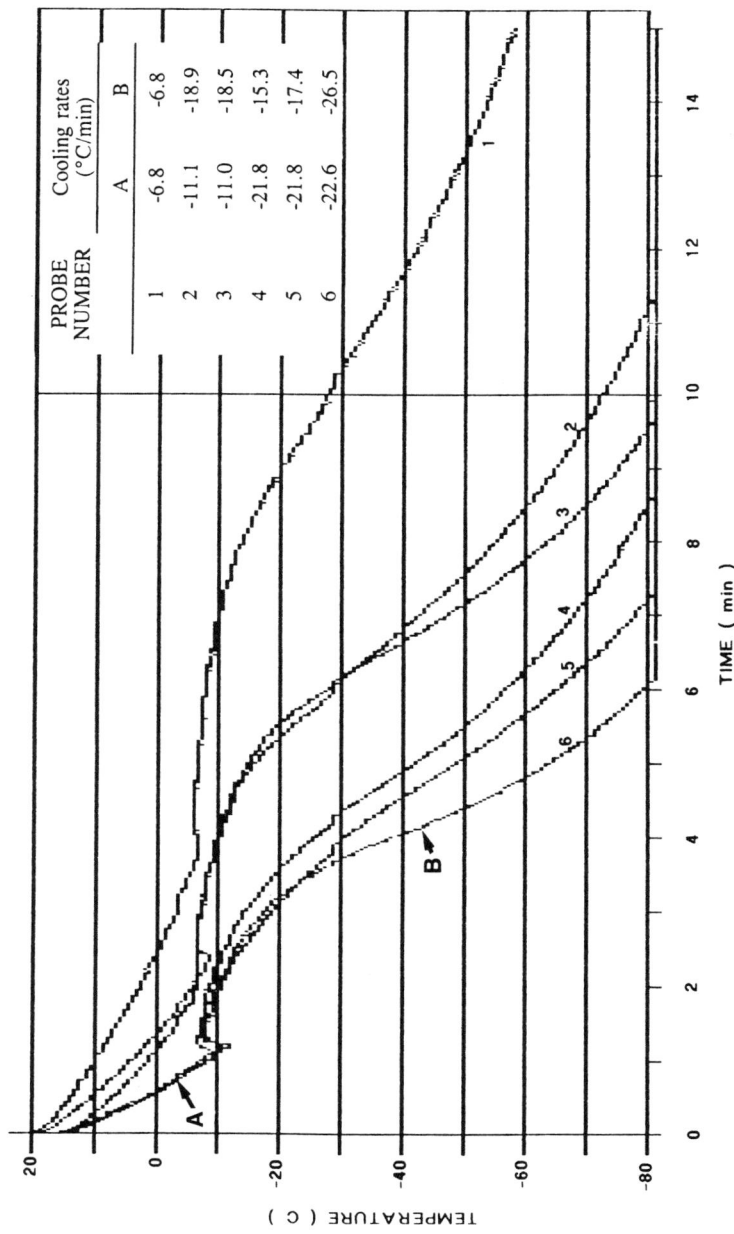

Fig. 1. Typical cooling profiles of the (**A**) pre- and (**B**) postfreezing phases within straws cooled in liquid nitrogen vapor phase in the neck of a Dewar. Cooling rates measured simultaneously in samples with a multichannel temperature logger.

physiology. Unlike mammals, the gametes of the vast majority of fish species are fertilized externally, their spermatozoa lack an acrosomal cap and have a comparatively short duration of motility. The spermatozoa remain quiescent in the tubules and spermduct of the testis and, in the case of freshwater fish, spermatozoa are activated by a reduction of external ionic strength of potassium and osmolality *(20,21)*. Moreover, fish spermatozoa are poor utilizers of external energy sources and the duration of motility of the majority of fresh water spermatozoa, unlike mammals, lasts only for between 15 s and 15 min *(4,5,22)*.

The reproducibility and efficacy of many published cryopreservation protocols remain unclear as many reports are incomplete. In several studies, the final spermatozoa concentration cryopreserved is not estimated or published, and the cooling rates are unknown, the volume of milt and the number of eggs used during insemination are not considered or given. Often, only positive results for protocols are published and therefore the true variability of protocols remain uncertain. To overcome these difficulties and to increase the use of cryopreservation technology, the methodology for fish spermatozoa cryopreservation should be standardized and minimum checks for evaluating protocols considered (*see* Notes 1–11).

Given the current state of the art of fish spermatozoa cryopreservation and species differences, one universal protocol cannot be offered. Instead, in this contribution, examples of successful protocols used for the cryopreservation of the four most widely studied groups of fish are presented in Table 1. For other protocols and fish groups the reader is directed to recent reviews on this subject *(3–6)*.

To develop reliable cryopreservation protocols for fish spermatozoa, emphasis should be placed on standardization. To contribute to this process of standardization, a flowchart outlining the stages and minimum quality control measures for cryopreservation are given in Fig. 2A,B, using a CRC and the Tilapia as models. The components of such a protocol is expanded in Sections 2. and 3., and supplement notes are provided for the major considerations in developing reproducible protocols.

2. Materials
2.1. Gamete Procurement

It is crucial that milt be collected in the quiescent form and should be free from contaminants, such as water, mucus, and gut exudate.

Table 1
Examples of Cryodiluents[a] Used for the Cryopreservation of Spermatozoa
of Key Groups of Fish of Aquacultural Importance

Constituents, g/L	Tilapia	Salmon	Trout	Carp
KCl	3.0	1.69	0.38	0.2
$CaCl_2$	0.3	0.14	—	0.2
NaCl	6.5	6.54	7.5	7.5
$NaHCO_3$	0.2	—	2.0	0.2
$MgSO_4 \cdot 7H_2O$	—	0.22	—	—
Glucose	—	—	1.0	—
Fructose	—	0.6	—	—
Glycine	—	6.0	—	—
Tris	—	—	24.2	—
BSA	—	—	4.0	—
Egg yolk (mL)	—	—	20	—
Promine-D	—	—	0–15	—
Distilled water (mL)	1000	1000	1000	1000
pH	8.6	6.4	7.3	—
Osmolality (mOsm/kg)	300	295	—	—
Cryoprotectant				
DMSO% (v/v)	—	10	10	15
MeOH% (v/v)	10	10	—	—
Success (% eyed eggs)	>90	20–80	20–80	55–85
Refs.	14,23	11,12	19	34

[a]A more comprehensive range of cryodiluents and protocols can be found elsewhere (3,6).

1. Fish should be starved for 6–24 h prior to milt collection. Ideally, spermatozoa should be collected during the natural spawning season although hormone therapy can also be selectively used to stimulate spermatogenesis, if required. The dose response for each candidate species should be researched (see Note 1).
2. For fertilization trials, freshly ovulated eggs should be pooled from at least three females. Females with ovulated eggs should be sedated and eggs stripped into clean dry containers and used within 1 h of collection.

2.2. Solutions

The composition of extenders used to dilute milt varies between fish species and studies. In this contribution, typical examples of widely used solutions for the salmonids, carp, and tilapia are presented Table 1 (see Note 2).

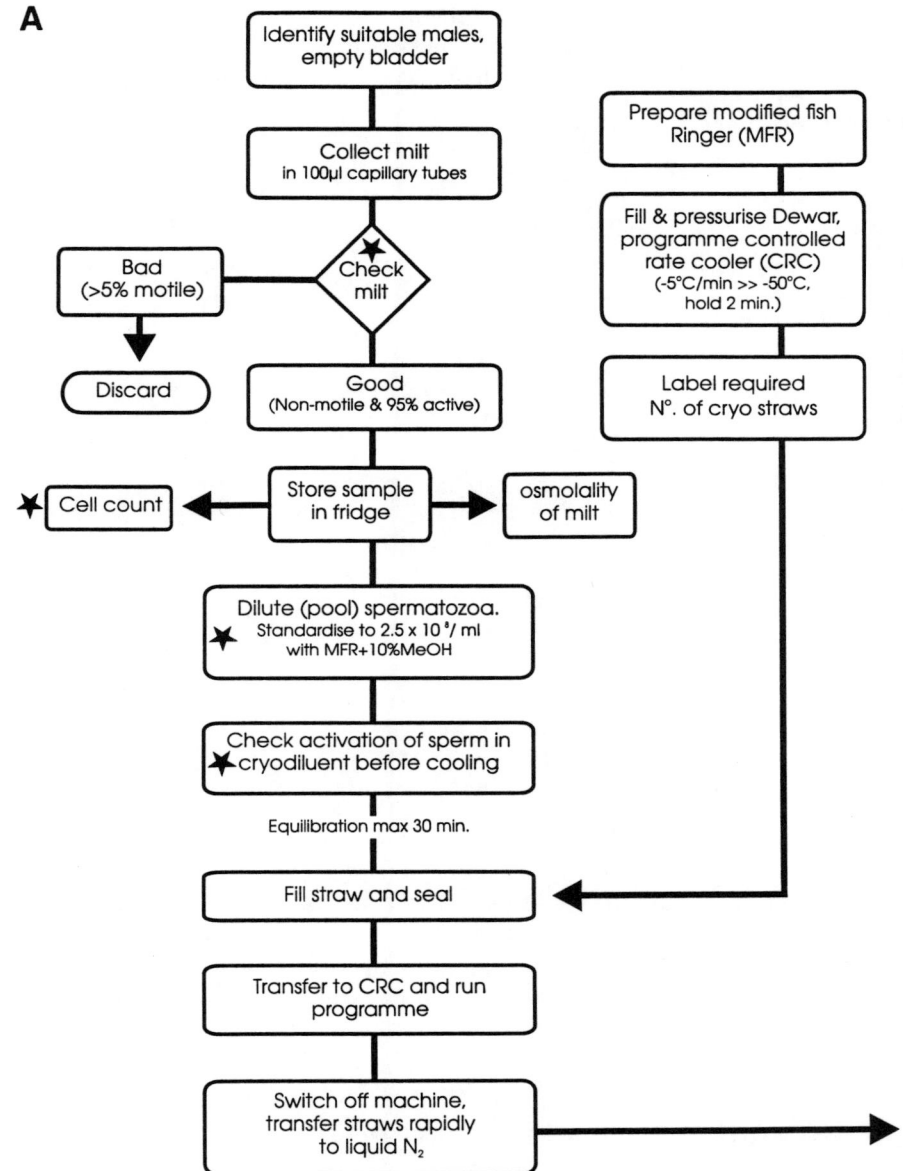

Fig. 2. Flowchart outlining the protocol for cryopreserving tilapia spermatozoa using a controlled rate freezer. Flow diagram showing (**A**) collection and cooling and (**B**) warming and

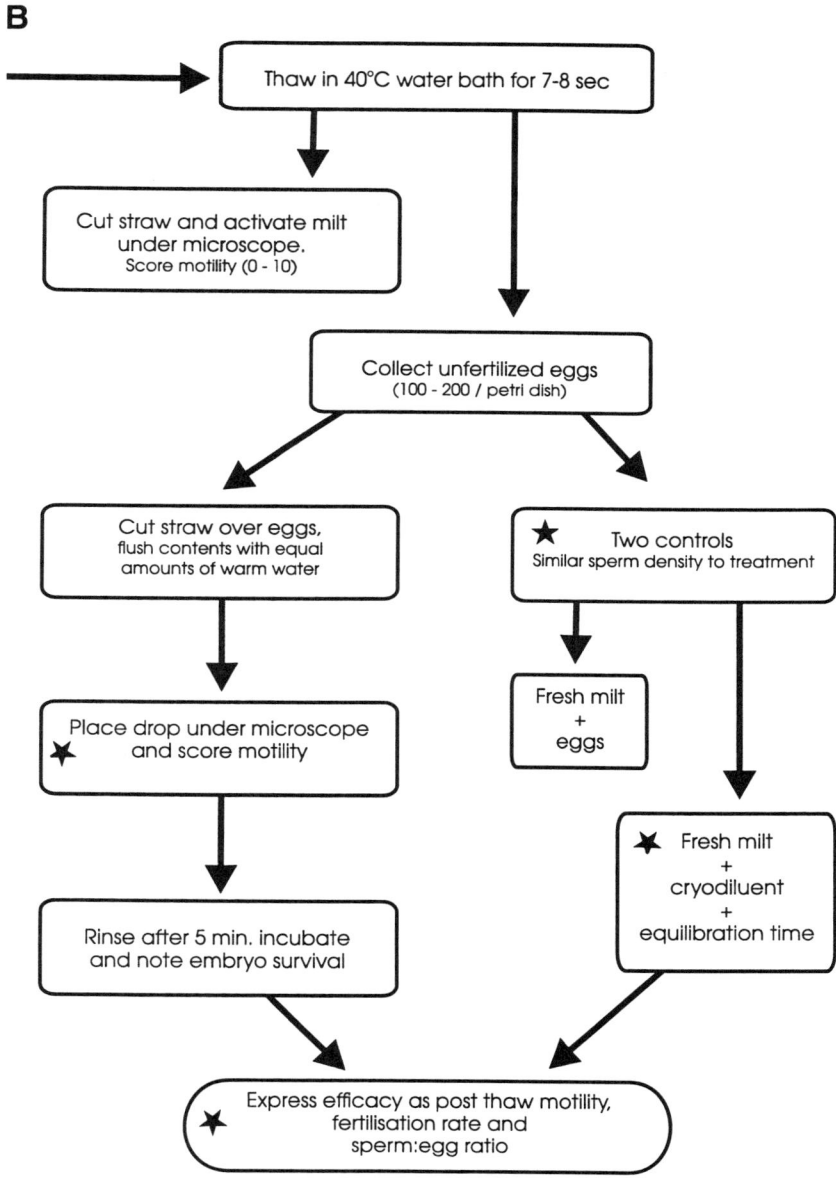

evaluation stages. Components regarded as essential for evaluating and standardizing protocols are indicated by (★) on the chart.

1. All solutions should be prepared in clean labeled glassware with distilled water and Analar grade chemicals. Extenders should be stored at 4–5°C and used within 1 wk. For longer term storage of solutions, aliquots can be stored in polypropylene containers at –20 to –70°C for several months. Prior to use, bottles should be thawed and shaken to aerate the solution and gently warmed if precipitation has occurred during storage.
2. Cryodiluents should be prepared prior to use and should be checked with a milt sample to ensure that spermatozoa are not activated by the diluent.
3. Deactivation solution: 0.6/0.7% (w/v) NaCl/KCl.

2.3. Equipment

1. Containers for milt collection: capillary tubes, sterile Eppendorf tubes, or tissue-culture flasks, and PVC powder or beads.
2. Programmable variable–rate cooler.
3. Safety equipment: cryogloves, goggles, visor, and so on.
4. Cryostorage vessels.
5. Water bath.
6. Hemocytometer.
7. Cryocontainers: straws or vials.
8. Polystyrene box.
9. Refrigerant: liquid nitrogen or dry-ice for alternative cooling method.

3. Method

1. Prepare containers for milt collection: The type of receptacle used for milt collection will depend on the size and spawning state of the fish. For small males, capillary tubes (10 –100 µL) should be used. For larger males, milt can be collected directly into sterile Eppendorfs or tissue-culture flasks.
2. Collect milt: Sedate males in spawning condition using benzocaine or MS-222, clear bladder, and express milt by abdominal massage into clean dry prelabeled containers (*see* Note 1).
3. Maintain milt samples at 4–5°C under aerobic conditions and use within 3 h.
4. Inspect a sample under a microscope (x100). Select samples showing no motility and >95% motility following activation for cryopreservation.
5. Estimate cell densities in a deactivator solution using a hemocytometer; densities in excess of 10^9/mL are adequate.
6. Establish optimum concentration of cryodiluent for each candidate species: For salmonoid, carp, and tilapia, make up final concentration of 5–10% (v/v) of appropriate cryoprotectant with chosen extender (*see* Table 1) (*see* Notes 2 and 3).
7. Milt dilution and equilibration: Dilute milt with appropriate cryodiluent. Establish optimum concentration of spermatozoa for each candidate spe-

cies. For tilapia, dilute milt with cryodiluent to give a final concentration of 2.5×10^8 cells/mL. This concentration is adequate to fertilize about 200 tilapia eggs (0.5/mL straw) (*see* Notes 3 and 4).
8. Load diluted milt into either prelabeled 0.25 or 0.5 mL straws or coded cryovials. For straws, allow an air space of 1–1.5 cm in straws to accommodate expansion of diluent during freezing.
9. Seal straws with either color-coded PVC powder or beads.
10. Equilibrate samples at 4–5°C for 20–30 min (*see* Note 4).
11. Following equilibration, load samples in the cooling apparatus for cryopreservation.
12. Cooling rates and the reproducibility of the protocol will depend on the apparatus used for cooling. Alternative methods are outlined in steps 12–14; all these have been used to successfully to cryopreserve fish spermatozoa (*see* Note 5).
    a. CRCs: Prior to use, fill liquid nitrogen dewar and bring dewar to working pressure. At the end of the program switch off the CRC to facilitate handling of samples (*see* Note 6).
    b. Establish optimal cooling rates for candidate species. For tilapia species, cool samples at 5°C/min to –50°C and hold 2 min.
    c. At the end of program switch off the machine and transfer the samples rapidly to liquid nitrogen for long–term storage (*see* Note 7).
13. Liquid nitrogen vapor cooling: Fill known volume or fix height of liquid nitrogen in a wide-necked dewar or polystyrene box (*see* Note 8).
    a. Monitor cooling profiles within several straws and establish cooling rates at various heights above the liquid nitrogen or empirically establish optimal height for cryopreservation of candidates fish species.
    b. Place samples on a rack at a predetermined height above the liquid nitrogen.
    c. Allow samples to cool for 10 min and then rapidly transfer to liquid nitrogen for long–term storage (*see* Note 7).
14. Dry-ice method (*see* Note 9): Predrill dry-ice block.
    a. Dispense 100–250-µL aliquots of cryodiluent containing milt into the predrilled holes.
    b. Allow to cool for 5 min.
    c. Remove and store frozen milt pellets in labeled cryovials under liquid nitrogen (*see* Note 7).
15. Warming procedure to recover frozen spermatozoa: Identify and remove desired sample and transfer rapidly to 40°C water bath. For 0.25–0.5-mL straws, thaw for 8 s. For trout pellets, drop the required number of pellets into $0.119M$ sodium bicarbonate solution until thawed and then add to eggs (*see* Note 10).

Table 2
Details of Cryopreservation Records for Fish Spermatozoa

Fish	Preservation details	Location
Species/strain	Date	No. of samples
Origin	Prefreezing quality	Color coding of straws and vistubes
Generation no.	Extender	Dewar no.
Identification no.	Cryoprotectant/conc	Canister position
Collection method /date	Dilution rate(s) /cell density	
Sperm density	Freeze method/cooling rate	
Milt volume	Postthaw quality	

16. Insemination and evaluation of protocols: Distribute predetermined quantities of freshly ovulated eggs into prelabeled dry containers (*see* Note 11).
    a. Establish two controls to evaluate egg quality and prefreezing and postthaw milt quality (*see* Note 5).
    b. Cut and empty the contents of straw over the eggs and activate spermatozoa with at least twice the volume of water at the normal rearing temperature and salinity of candidate species.
    c. Place a drop of fluid from the egg/sperm mixture on a glass slide and score motility on scale of 0–10; 0, no activation and 10, total activation of spermatozoa.
    d. For postthaw motility evaluation, place and activate postthawed milt sample on a slide and note motility score following activation as in step 16c.
17. Record keeping: Maintain comprehensive records of cryopreserved material. The basic information that should be recorded is given in Table 2.

4. Notes

1. Milt collection and quality: The outcome of cryopreservation will depend on the quality of the prefrozen fish milt. The structure and cryoresistance of spermatozoa and the physicochemical composition of seminal plasma varies between individuals and species *(5,8)*. Therefore the prefreezing quality of milt of candidate species should be carefully researched. The quality of spermatozoa even of the same species varies considerably. Fitness of spermatozoa may vary with prefreezing storage time and conditions *(23)*, with season *(24–27)*, social hierarchy *(7)*, diet *(16)*, collection techniques *(8)*, and need to be carefully researched for each species. To

maximize the success of cryopreservation of milt from seasonal spawners, milt should be collected during the middle of the breeding season *(24)*. Milt from individual fish should be held in isolation until screened for quality, and then pooled, if required.

To maximize milt collection, broodstock should be maintained under stress-free conditions. Large species, such as carp, catfish, and salmon should be sedated to facilitate handling and minimize damage to personnel and fish. Anesthetic dose required sedation for different fish sizes and species should be researched. For tropical species, such as tilapia, broodfish should be maintained with females in a spawning environment to stimulate milt production. In addition to the spawning state, the ease of milt collection and quality will vary with the quantity of available milt and size of the males. The risk of urine and fecal contamination of milt from males producing small quantity of milt is greater.

Milt contaminated with mucus, water, and feces will contain activated spermatozoa. To minimize contamination, testis of large species can be catheterized or as in most species the contaminants expelled by abdominal massage prior to collection of milt. Milt should be stored in containers offering the largest possible surface area to facilitate oxygenation but condensation within storage vessels should be avoided. It is generally suggested that milt should be frozen as soon as possible after collection *(3,19,28)*. For tilapia, milt can be successfully frozen following 6 d of storage at 4°C *(23)*. For Atlantic salmon *(Salmo salar)*, prefreezing storage time can be extended to 21 d (unpublished data).

2. Extenders: Comprehensive reviews on fish extenders can be found elsewhere *(3,6)*. To conserve the finite endogenous energy reserves of fish spermatozoa, extenders should be tailored to maintain the spermatozoa completely quiescent. Successful cryopreservation does not correlate with the complexity of the extenders *(3)*. The total ionic strength appears to be of major importance and simple sugar solutions with osmolalities similar to seminal plasma have been successfully used to cryopreserve trout milt *(10,19)*.

3. Cryoprotectants: Efficacy of cryprotectants will depend on the prefreezing toxicity of compounds to spermatozoa *(14)*, concentration, equilibration period, and the extender used *(23,28)*. Cryoadditives, such as ethylene glycol *(29)*, glycerol *(29–31)*, DMSO *(29,31–36)*, MeOH *(14,23,28)*, and dimethyl acetamide *(32)* have been successfully used for fish spermatozoa cryopreservation. In most studies glycerol has been demonstrated to be inferior to DMSO and MeOH. Generally, concentrations of 5–15% (v/v) of either DMSO or MeOH have been successfully used for most fish species *(3–6)*. The choice of cryoprotectant may finally depend on availability

of cryoprotectant, facilities, and toxic hazard considerations to users and fish spermatozoa. DMSO is generally regarded to be highly toxic to humans, and, ideally, requires a fume cupboard and nitrile gloves for safe handling. MeOH is relatively cheap, readily available, and is less hazardous to users.

Nonpermeating protectants, such as milk powder *(28,31,36)*, egg yolk *(15,31,36)*, and proteins *(9)*, have been used, but their importance for candidate species need to evaluated under standardized conditions.

4. Dilution and equilibration time: For salmonoid, carp, and tilapia, milt is generally diluted 3–20-fold. To establish the efficacy of protocols and to compare protocols between studies, this variable must be standardized. Since sperm density shows high variation within and between individuals, between species, and during breeding season, the use of dilution ratio must be complimented with cell density. The final number of spermatozoa/mL of diluted milt and not dilution ratio, should be established and standardized.

   Prior to or during cooling, a sample of spare prefrozen milt should be inspected under a microscope (x100) and prefreezing motility assessed on a scale of 0–10.

   Although it is generally suggested that equilibration time is not necessary, in practice, depending on the cooling method, number of treatments and samples being processed, equilibration times of 20–25 min are common. The contact time between spermatozoa and cryodiluents before cooling should not exceed 30 min.

5. Cooling apparatus: The cooling method used for fish spermatozoa cryopreservation will depend on the acceptable level of reproducibility of the technique and the purpose of cryopreservation. The reproducibility of cooling rates and maximal fertilization rate should be established for each method. The cooling characteristics within storage containers during the cooling process will vary considerably, depending on the method and protocol used.

6. For reproducible results, a CRC is ideal and is particularly useful for establishing optimal cooling rates for spermatozoa of different fish species. The cooling rates within straws in CRC can vary depending on the size of the straws or cryovials, loading of the chamber, and the position of the sample in the cooling chamber.

7. Storage: Ideally, individual fish spermatozoa samples should be retrievable with minimal disturbance to other straws. Storage in wide-necked refrigerators can increase the risk of thawing during sample location and retrieval *(14)*, and this system should be avoided for fish spermatozoa.

8. The cooling rates generated using liquid nitrogen vapor column will vary with the type of cooling apparatus used. In a polystyrene box, for example,

the pre- and postfreezing cooling rates can vary by as much as 30°C, and the cooling rates between different straws at any given height can be variable. This variability is greater when samples are cooled in the neck of the dewar (Fig. 1)

9. The dry-ice cooling method offers only one freezing rate of 30–35°C/min *(3)* and has been widely used for trout *(9,10,18,19,33)*. This method, though attractive for field use, may be too fast for some species and therefore may be of limited use. In comparative trials, however, postthaw viability of spermatozoa cryopreserved in straws were reported to be more superior to those cryopreserved as pellets *(29)*. In addition, pellets cannot be hermetically sealed and fish pellets are prone to crumbling.

10. Warming procedure: A visor or safety goggles must be worn during thawing. To facilitate later identification of straws used for motility scores, they should be cut at the cotton plug end.

11. Insemination and protocol evaluation: During fertilization trials it is imperative to evaluate the motility of the prefrozen diluted milt and postthawed milt in the egg and milt mixture. Following insemination, a sample of ovarian fluid containing the milt should be immediately sampled and scanned under a microscope to establish the level of sperm activity. In some cases the eggs should be quickly rinsed to flush away the contents of broken eggs that may inhibit sperm activation. Fertilization success should be established after the late blastula stage, preferably at the eyed or later stages. In all cases the number of spermatozoa/egg used for the insemination trials must be estimated.

To evaluate protocols, ideally, two controls should be established. Fresh milt should be used at a similar sperm concentration to the postthawed sample to establish the quality of eggs. In addition, to separate the effects of pre and postfreezing events, a sample of fresh spermatozoa at a similar density should be equilibrated for a similar period in the same cryodiluent as the postthawed milt and used to fertilize eggs.

Acknowledgment

The data presented in this chapter were gathered under grants and facilities provided by the British Overseas Administration at the Institute of Aquaculture, University of Stirling, Scotland.

References

1. Polge, C., Smith, A. U., and Parkes, A. S. (1949) Revival of spermatozoa after vitrification and dehydration at low temperature. *Nature* **164,** 166.
2. Blaxter, J. H. S. (1953) Sperm storage and cross fertilisation of spring and autumn spawning Herring. *Nature* **172,** 1189,1190.

3. Scott, A. P. and Baynes, S. M. (1980) A review of the biology, handling and storage of salmonoid spermatozoa. *J. Fish Biol.* **17,** 707–739.
4. Stoss, J. (1983) Fish gamete preservation and spermatozoa physiology fish physiology. *J. Fish Biol.* **22,** 351–361.
5. Leung, L. K.-P. and Jamieson, B. G. M. (1991) Live preservation of fish gametes, in *Fish Evolution and Systematics: Evidence from Spermatozoa* (Jamieson, B. G. M., ed.), Cambridge University Press, Cambridge, UK, pp. 245–269.
6. McAndrew, B. J., Rana, K. J., and Penman, D. J. (1993) Conservation and preservation of genetic variation in aquatic organisms, in *Recent Advances in Aquaculture* (Muir, J. F. and Robert, R., eds.), Blackwell Scientific Publications, Oxford, UK, pp. 295–336.
7. Rana, K. J. (1993) Cryopreservation of fish spermatozoa, in *Workshop Proceedings: Genetics in Aquaculture and Fisheries Management* (Penman, D., Roongratri, N., and McAndrew, B., eds.), 31st August–4th September 1992, University of Stirling, Stirling, UK, pp. 49–53.
8. Rana, K. J (in press) In vitro preservation of fish gametes, in *Broodstock Management and Egg and Larval Quality* (Bromage, N. R and Roberts, R. J., eds.).
9. Stoss, J. and Holtz, W. (1983) Cryopreservation of rainbow trout milt (*Salmo gairdneri*). III. Effect of proteins in the diluents, sperm from different males and interval between collection on freezing. *Aquaculture*, **31,** 275–282.
10. Holtz, W., Schmidt-Baulain, R., and Meiners-Gefkin (1991) *A Simple Saccharide Extender For the Cryopreservation of Rainbow Trout* (Oncrhynchus mykiss). Proceedings of the Fourth International Symposium on the Reproductive Physiology of Fish, University of East Anglia, Norwich, UK, pp. 250–252.
11. Hoyle, R. J. and Idler, D. R. (1968) Preliminary results in the fertilisation of eggs with frozen sperm of Atlantic salmon *(Salmo salar). J. Fish Res. Bd. Can.* **25,** 1295–1297.
12. Ott, A. G. and Horton, H. F. (1971) Fertilisation of steelhead trout (*Salmo gairdneri*) eggs with cryopreserved sperm. *J. Fish Res. Bd. Can.* **28,** 1915–1918.
13. Ott, A. G. and Horton, H. F. (1971) Fertilisation of chinook salmon eggs with cryopreserved sperm. *J. Fish Res. Bd. Can.* **28,** 745–748.
14. Rana, K. J. and McAndrew, B. J. (1989) The viability of cryopreserved Tilapia spermatozoa. *Aquaculture* **76,** 335–345.
15. Baynes, S. M. and Scott, A. P. (1987) Cryopreservation of Rainbow trout spermatozoa: the influence of sperm quality, egg quality and extender composition on post thaw fertility. *Aquaculture* **66,** 53–67.
16. Erdhal, A. W. (1986) *Factors Affecting Storage and Fertilisation of Fish Gametes.* Thesis submitted to the faculty of the graduate school of University of Minnesota.
17. Chao, H. N., Tsai, C. T., Hsu, H. W., and Linn, T. T. (1987) The properties of Tilapia sperm and its cryopreservation. *J. Fish Biol.* **30,** 107–118.
18. Schmidt-Baulain, R. and Holtz, W. (1989) Deep freezing of Rainbow trout sperm at varying intervals after collection. *Theriogenology* **32,** 439–443.
19. Stein, H. and Bayrle, H. (1978) Cryopreservation of the sperm of some freshwater teleosts. *Ann. Biol. Anim. Biophys.* **18,** 1073–1076.

20. Morisawa, M., Suzuki, K., and Morisawa, S. (1983) Effects of potassium and osmolality on spermatozoan motility of salmonoid fishes. *J. Exp. Biol.* **107,** 105–113.
21. Morisawa, M. and Suzuki, K. (1980) Osmolality and potassium ion: their roles in initiation of sperm motility in teleosts. *Science* **210,** 1145,1146.
22. Billard, R. (1988) Artificial insemination and gamete management in fish. *Marine Behav. Physiol.* **14,** 3–21.
23. Rana, K. J., Muiruri, B., and McAndrew, B. J. (1990) The influence of equilibration time, and pre-freezing storage time, on the viability of cryopreserved *Oreochromis niloticus* (L) spermatozoa. *Aquaculture Fish. Manage.* **21,** 25–30.
24. Benau, D. and Terner, C. (1980) Initiation prolongation and reactivation of the motility of salmonoid spermatozoa. *Gamete Res.* **3,** 247–257.
25. Piironen, J. and Hyvarinen, H. (1983) Composition of the milt of some teleost fishes. *J. Fish Biol.* **3,** 351–362.
26. Munkittrick, K. R. and Moccia, R. D. (1987) Seasonal changes in the quality of rainbow trout (*Salmo gairdneri*) semen: effect of delay in stripping on spermatocrit, motility volume and seminal plasma constituents. *Aquaculture* **64,** 147–156.
27. Kruger, J. C., DeW., Smit, G. L., Van Vuren, J. H. J., and Ferreira, J. T. (1984) Some chemical and physical characteristics of the semen of *Cyprinus carpio* L and *Oreochromis mossambicus* (Peters). *J. Fish Biol.* **24,** 263–272.
28. Harvey, B. (1983) Cryopreservation of *Sarotherodon mossambicus* spermatozoa. *Aquaculture* **32,** 313–320.
29. Erdhal, A. W., Erdhal, D. A., and Graham, E. F. (1984) Some factors affecting the preservation of salmonoid spermatozoa. *Aquaculture* **43,** 341–350.
30. Steyn, G. S. and Van Vuren, J. (1987) The fertilising capacity of cryopreserved sharptooth catfish (*Clarias gariepinus*) sperm. *Aquaculture* **63,** 187–193.
31. Leung, L. K. P. (1987) Cryopreservation of spermatozoa of the barramundi, *Lates calcarifer* (Teleostei: Centropomidae). *Aquaculture* **64,** 243–247.
32. Gallant, R. K. and McNiven, M. A. (1991) Cryopreservation of rainbow trout spermatozoa. *Bull. Aqua. Assoc. Can.* **3,** 25–27.
33. Stoss, J. and Retstie, T. (1983) Short term storage and cryopreservation of milt from Atlantic salmon and sea trout. *Aquaculture* **30,** 229–236.
34. Kurokura, H., Hirano, R., Tomita, M., and Iwahashi, M. (1984) Cryopreservation of carp sperm. *Aquaculture* **37,** 267–273.
35. Steyn, G. J. and Van Vuren, J. H. J. (1991) Cryopreservation of the spermatozoa of 2 African freshwater fishes (*Characidae*). *S. African Tydsker Natuur..avo* **21,** 76–81.
36. Chao, H. N., Chao, W.-C., Liu, K.-C., and Liao, L.-C. (1987) The properties of Tilapia sperm and its cryopreservation. *J. Fish Biol.* **30,** 107–118.

Chapter 17

Cryopreservation of Avian Spermatozoa

Graham J. Wishart

1. Introduction

Modern animal and plant cell cryopreservation technology began when Polge and colleagues demonstrated the cryoprotective action of glycerol by producing live chicks from domestic fowl hens inseminated with spermatozoa that had been frozen to −79°C *(1)*. Since then there have been many reports extending these findings to other species of poultry and nondomesticated birds; describing alternative freezing/thawing protocols, cryopreservatives, diluents, and insemination regimes; and demonstrating or proposing applications of the technique to the poultry industry (*see* refs. *2–6* for reviews).

Thus spermatozoa from turkey, duck, goose *(3,5)*, American kestrel *(7)*, sandhill crane *(8)*, peregrine falcon *(9)*, and budgerigar *(10)* have all been shown to produce viable offspring after freezing to −196°C. However, by far the most studies have been performed with the domestic chicken and the method to be described here has been designed primarily for that species, although it has been used in a similar form for other species.

Glycerol continues to be widely used as a cryoprotectant for avian spermatozoa, although it has a contraceptive action that is manifested within the hen's vagina. Glycerol must therefore be removed from sperm suspensions by dialysis, or by centrifugation and resuspension of spermatozoa in glycerol-free medium before intravaginal insemination *(5)*. Alternatively, fertility can be obtained from glycerol-semen mixtures by circumventing the vagina and inseminating into the uterus, magnum, or infundibulum *(5)*.

Other cryoprotectants that have been used for freezing chicken spermatozoa and that do not have an obvious contraceptive effect are not generally removed before insemination. Examples of these include: dimethyl sulfoxide, dimethylacetamide, dimethylformamide, ethanediol, propanediol, and methylpyrollidone *(4,5)*. However, in studies that have compared the fertility obtained from hens inseminated with different cryoprotectants, the "best" fertility has generally been obtained with glycerol *(5)*. Thus, despite the additional technical manipulations involved in removing glycerol before insemination by the more convenient vaginal route, it is employed in the recommended method described here.

The nature of the damage inflicted on chicken spermatozoa by cryopreservation is, in many respects, unclear. In vitro studies, which show that 30–40% of spermatozoa survive with their functional and metabolic integrity intact, grossly overestimate their retention of fertilizing ability *(11)*. This has been reported as 1.6% of that of untreated spermatozoa *(11)* and 19.7% of that of glycerolized, centrifuged spermatozoa *(12);* the difference in the two figures being, at least partly, accounted for by the additional manipulation. It is thus unfortunate that in vitro assays of sperm "quality" that can be used to predict fertility differences between samples of semen from different male birds, are not useful for predicting the fertility of cryopreserved spermatozoa *(11)*, and that an appropriate assessment of a particular protocol must be made by the time-consuming fertility trial. This generally involves inseminating each of a group of hens with a given dose of spermatozoa and obtaining an estimation of either the mean percentage of fertile eggs laid by the hens during 1 or 2 wk after insemination, or the median number of days postinsemination between the laying of the first infertile and last fertile egg *(13)*. Chicken hens are extremely variable in their response to inseminated spermatozoa so that groups of at least 15–20 birds are required to obtain a reasonable estimation of the fertility of a sample of semen.

Using such information to select a protocol, based on the "best" hen fertility achieved with chicken spermatozoa frozen in glycerol, requires a thorough examination of all stages of each technique in order to avoid misapprehension.

For example, new methods are often presented without concurrent experimental comparison to an established method and, since different strains of birds will have variable fecundity *(14,15)*, the resultant fertility levels are, in comparative terms, arbitrary. Even before and after freezing comparisons of fertility can be misleading if the insemination dose is not carefully controlled: When >90% of fertile eggs are produced by hens inseminated with fresh semen, their fertility will be poorly related to the insemination dose *(11)* and will not represent a suitable control for cryopreserved samples. In addition, some reports are covertly incomplete: For example, the depth of insemination into the vagina can have a considerable effect on the fertility induced, and is not always specified.

The method described in the following sections is essentially that devised by Lake and colleagues for chicken spermatozoa *(16,17)*. The method, most of which is carried out in a 4–6°C cold room, employs a programmable freezer and utilizes glycerol as cryoprotectant so that a refrigerated centrifuge is also required to wash the cells free of cryoprotectant before insemination.

Although this method involves the use of equipment that may not be readily available in a laboratory that is not routinely involved in cryopreservation, its choice is justified by the consistently high fertility results obtained with it both in the laboratory *(16)* and in the field *(17)*, and when performed by different operators *(14,15)*, on spermatozoa from domestic fowl. Furthermore, it has proven useful for cryopreservation of spermatozoa from other species *(7,9,10,17)*. Alternative materials or manipulations that may be more convenient under different situations are listed as Notes and referred to throughout Sections 2. and 3.

2. Materials

2.1. Semen

It is important that semen should be collected as free as possible from contaminant feces or urates from the cloaca. Whether or not "transparent fluid," the exudate from cloacal glands, is detrimental or beneficial to sperm function remains speculative, but it certainly dilutes the sperm concentration of semen samples and on that basis it should be avoided. A concentration of $3-5 \times 10^9$/mL is generally considered to be normal. Ideally birds providing semen should be placed on a routine collection regime, for example, twice weekly. This will familiarize both birds and operator with the technique and thus help maximize the quantity of semen obtained and minimize its contamination. It will also induce a regular flow of semen through the ductus deferens, enabling the collection of fresher, better quality, spermatozoa. Whether kept as samples from individual males or pooled, the semen should be cooled to 5°C as soon as possible after collection for processing within 15 min.

Housing and husbandry are very important for maintaining the birds in good reproductive condition and information or advice should sought on this from relevant sources. These, and semen collection techniques, will vary between species and should be researched carefully.

2.2. Solutions

These should be prepared using distilled or deionized water and all glassware should be thoroughly cleaned and rinsed before use. Clean but not sterile preparative techniques are recommended, although if solutions are to be kept under "field" conditions, sterilizing them by filtration would be a worthwhile precaution. Chemicals are analytical reagent grade. Glycerol should be stored at 4°C. Both diluent (step 3) and cryoprotectant (step 4) solutions can be made on the day of use or stored at 4°C for a few days. Alternatively, they can be transferred in 20-mL aliquots to plastic containers and stored at −20°C for several months.

1. Cryoprotectant solutions (see Table 1): The glycerol should be equilibrated at room temperature before use to reduce its viscosity and thus facilitate pipeting. Even then its viscosity is too great to permit accurate pipeting so that quantities must be measured by weight. Glycerol and all solid materials may be placed in the same container and dissolved together.
2. Diluent (see Table 2): All constituents may be placed in the same container and dissolved together (see Note 3).

2.3. Major Equipment

1. Cold room maintained at 4–6°C (see Note 1).
2. Refrigerated centrifuge capable of running at 5°C.
3. Programmable freezer: This should be programmed to hold at 5°C until loaded with samples, run at −3°C/min from 5 to −35°C, and then to hold the samples at this temperature for 10 min (see Note 4).
4. Liquid nitrogen storage container, with cylinder-type containers for holding canes.

2.4. Minor Equipment

1. Positive displacement adjustable pipet (0.1–1.0 mL). Both semen and the cryoprotectant solution are viscous so that volumes dispensed with an adjustable air-displacement pipet are unreliable. As an alternative, capillaries or glass pipets may be used.

Table 1
Cryoprotectant Solution

Constituent	M	g/100 mL
Glycerol	1.480	13.64
Sodium glutamate (monohydrate)	0.103	1.92
Magnesium acetate (tetrahydrate)	0.004	0.08
Potassium acetate (anhydrous)	0.051	0.50
Polyvinyl pyrrolidine (MW 10,000)	0.0003	0.30
Glucose	0.044	0.80

Table 2
Diluent

Constituent	M	g/100 mL
Sodium glutamate (monohydrate)	0.103	1.92
Magnesium acetate (tetrahydrate)	0.004	0.08
Tripotassium citrate (monohydrate)	0.004	0.13
Sodium acetate (anhydrous)	0.062	0.51
Glucose	0.033	0.60

2. 2-mL Cryotubes (*see* Note 2).
3. Aluminum canes for holding vials: A small lead weight ($\approx$ 8 g) attached to the lower end of these will prevent the canes with partially filled cryotubes floating in liquid nitrogen or aqueous solutions.
4. Water bath precooled to 5°C (*see* Note 1).
5. Thawing bath precooled to 5°C: This consists of a 1 L volume polypropylene measuring cylinder filled with a 20% aqueous solution of ethylene glycol (or ethanol) at 5°C.

3. Method

3.1. Freezing Protocol (see *Note 4*)

All manipulations should be performed in a cold room, if available. Otherwise samples should be kept in water baths capable of retaining a temperature of 5°C. Before starting the following steps, the programmable freezer should be made ready with the liquid nitrogen container attached and pressurized.

1. Bring semen to 5°C and dispense 0.15 mL into cryovials held in the 5°C bath (*see* Note 2).

2. Add 0.45 mL of cryoprotectant solution, already equilibrated to 5°C, to each vial. Fix caps, mix thoroughly by agitation, and replace into the 5°C bath. Equilibrate for 10 min at this temperature. The chamber of the programmable freezer with canes in position may be brought to 5°C during this time.
3. Transfer vials quickly to their canes within the freezer chamber.
4. Run program cooling samples from 5 to −35°C at a cooling rate of 3°C/min (*see* Note 4). Allow at least 5 min at −35°C to ensure that all samples have reached the end temperature. (This may vary at different locations within the freezer chamber.)
5. Transfer the canes with attached vials into the liquid nitrogen storage container, immersing quickly to avoid warming of samples. Wear eye protection and insulating gloves during this procedure. Store the frozen material until it is required.

3.2. Thawing Procedure

As for the freezing procedures, the following steps should be carried out in a 5°C cold room, if available. The refrigerated centrifuge should also be equilibrated at 5°C.

1. Transfer the canes with attached vials from the liquid nitrogen storage container to the 5°C thawing bath. It is most important to wear hand and eye protection during this stage and to hold the canes at arms length before dropping them into the bath, so that any violent vaporization of nitrogen is not dangerous.
2. Agitate the canes to expedite thawing and disperse any ice that may form around the vials. When the samples are seen to thaw, transfer them to the 5°C benchtop water bath.
3. Dry any liquid adhering to the neck of the vial and remove the screwcap. Using a Pasteur pipet, transfer the contents of the vial to a 10–15 mL centrifuge tube, already equilibrated to 5°C in the bath.
4. Add diluent, at 5°C, to each tube by the following staged additions of: 0.08, 0.22, 0.40, 0.73, 1.50, 1.90-mL vol. Mix well by gently aspirating with a Pasteur pipet after each addition and allow 2–3 min in between additions (*see* Notes 3 and 5).
5. Transfer the tubes to a refrigerated centrifuge, precooled to 5°C, and centrifuge at 700g for 15 min.
6. Remove tubes from the centrifuge, one by one, and decant the supernatant by carefully inverting the tube, removing excess liquid around the rim of the tube with a Pasteur pipet or filter paper. Replace the righted tubes in the 5°C bath. Any fluid that drains from the side of the tube back onto the

pellet should be removed quickly. Active spermatozoa will swim up from the pellet into this excess fluid, even at 5°C, but especially at higher temperatures. Should active spermatozoa swim up into the excess liquid, be aware that it is more important to remove the glycerol than to "save" spermatozoa; excess glycerol will reduce the fertility of all remaining spermatozoa within the sample.
7. Add 0.1 mL of diluent, at 5°C, to each tube and mix by agitation, flicking the end of the tube. A final, gentle, aspiration with a Pasteur pipet will ensure complete resuspension.

3.3. Artificial Insemination

This is a critical stage; poor insemination technique will negate the efforts made in carrying out these protocols. Artificial insemination of semen samples into the hen is usually performed intravaginally *(13)*. This involves everting the cloaca by applying digital pressure to the hen's abdomen, which should expose the vaginal opening. Further, gentle, pressure will reveal this more distinctly and evert about 1 cm of the distal vagina. Insemination of $2-6 \times 10^8$ spermatozoa in a volume of, ideally, no more than 0.15 mL should be made about 3–4 cm into the vaginal opening. Expulsion of spermatozoa may be reduced by relaxing abdominal pressure just before insemination. The frozen/thawed samples of semen should be inseminated as soon as possible after reconstitution, certainly within 20 min. During this time the semen should be held at 5°C, although it has been recommended that samples be warmed by exposure to room temperature 30 s before insemination *(16)*. A series of inseminations, given at intervals of a few days, will improve the maintenance of hen fertility *(17,18)*. Inseminations should, ideally, be made about 4–8 h after oviposition (*see* Note 6).

3.4. Assessment of Fertility

Fertile eggs will, at least potentially, be laid between 40–44 h after insemination. These may be tested for fertility by "candling" with a light source after about 7 d of incubation. Eggs that remain clear after 10 d should be opened to check for signs of embryonic development so that primary fertility can be distinguished from early embryonic death (*see* Note 7).

4. Notes

1. Ideally, all procedures detailed should be carried out in a cold room. If this is not available, then a "makeshift" alternative can be constructed *(17)*.

However, it is possible to maintain samples at 5°C in the laboratory by the use of water baths. Under such conditions, the sample manipulation and transfer must be made quickly and efficiently. Also, pipets and pipet tips should be held in diluent within the baths to maintain 5°C throughout.

Whatever the conditions, bath temperature should be monitored and should not exceed 8°C. It is also important to note that spermatozoa may be damaged if the bath temperature falls below 4°C.

2. Alternative containers for frozen semen: In most of the published work using Lake's method, semen has been frozen in 1-mL glass ampules that are flame-sealed after mixing the semen and cryoprotectant. The temperature of the body of the ampule and its contents are maintained at 4–5°C by immersing them in water at 5°C during sealing *(16,17)*. Such a process is technically involved and inconvenient. Semen frozen in 0.5-mL plastic straws did not show any significant difference in resultant hen fertility when compared to the ampule method *(19)*. The cryotubes recommended in this procedure are much more convenient to use and are designed for efficient heat transfer. However, it should be noted that they have not been compared with glass ampules or straws in extensive fertility trials. A new alternative receptacle for freezing semen is the "CryoCell." This has, contained within a plastic frame, porous membranes that are impervious to glycerol during freezing and storage, but rendered permeable during the thawing process so that glycerol is removed by dialysis *(18)*.

3. Alternative diluent and cryoprotectant solutions: Although the various alternative solutions for avian sperm cryopreservation may be found by consulting the reports cited in Section 1., two glycerol-based solutions are in current use for cryopreservation of chicken spermatozoa and are therefore worth mentioning here. The Minnesota Avian buffer is used as a diluent and, with the addition of 8% (v/v) glycerol, as a cryoprotectant solution *(20)*. These have had a degree of success in fertility trials *(21)*.

The Hiroshima diluent is available commercially as solution A (diluent) and solution B (cryoprotectant) from the Kyoritsu Shoji Company (Tokyo, Japan). Fertility levels using these solutions in a pellet cryopreservation protocol have been reported to be greater than 70% following weekly inseminations *(21)*.

4. Alternative freezing protocols: Several other workers use a programmable freezer, generally with a starting temperature of 5–6°C, a freezing rate of –3 to –6°C/min, and a "plunge" temperature of –40 to –45°C *(14,18,20)*. Japanese researchers have reported a method involving freezing samples in the vapor phase above liquid nitrogen before plunging *(22)* and also

freezing in pellets made by dropping semen–cryoprotectant mixtures onto depressions made in solid carbon dioxide before transferring to liquid nitrogen *(21,23)*.
5. Alternative dilution protocol: If several duplicate samples are being processed, then it may be more convenient to pool them and dilute them automatically. This is achieved by adding diluent continuously from a peristaltic pump into a semen sample that is mixed gently with a magnetic stirrer. A diluent volume of eight times that of the cryopreserved sample should be added over 15 min. The diluted samples should be subsequently centrifuged as 5-mL aliquots.
6. Alternative insemination procedures: Intravaginal insemination is likely to be the preferred route for large-scale trials and field applications. However, most inseminated spermatozoa are lost during their transport through the vagina, so that inseminations performed beyond this site will enable more spermatozoa to colonize the oviduct. This may be achieved by intrauterine (IU) insemination, in which the uterus is palpated digitally and breached, allowing the insemination tube to be placed directly into the uterus. The main problem with this technique is that breeching of the uterovaginal junction tends to interrupt the hen's laying cycle for at least a few days. A recent report has described an improved IU insemination protocol *(24)*.

   Insemination into the magnum is performed by injecting samples into that region of the oviduct after its exposure during surgery under anesthesia. Clearly, such measures should not be considered lightly. However, they produce high fertility with few and poor quality spermatozoa *(25)*.
7. Alternative ways of assessing fertility: The most obvious way to assess fertility is to count the proportion of fertile eggs laid by inseminated hens. If this is done after inseminating a range of doses of spermatozoa, then a more quantitative assessment of the effects of cryopreservation on sperm fertilizing ability may be achieved *(11)*. Alternatively, inseminations in which spermatozoa from two phenotypically distinct strains of chickens compete for fertilization may be used to compare fertilizing ability before and after cryopreservation *(20)*. Also, since inseminated spermatozoa are found in the perivitelline layer of laid eggs in numbers that are proportional to the dose of viable spermatozoa inseminated *(11)*, these may be used to assess the success of a cryopreservation protocol. Although the fertilizing ability of spermatozoa cannot yet be predicted from tests performed in vitro, an assessment of morphology and "live/dead" status of spermatozoa after staining with nigrosin and eosin *(13)* may provide a useful measure of gross damage to spermatozoa.

Acknowledgment

I would like to thank Peter Lake for his advice during the preparation of the manuscript.

References

1. Polge, C. (1951) Functional survival of fowl spermatozoa after freezing at −70°C. *Nature* **167,** 949,950.
2. Sexton, T. J. (1983) Maximising the utilisation of the male breeder: a review. *Poult. Sci.* **62,** 1700–1710.
3. Graham, E. F., Schmehl, M. L., and Deyo, R. C. M. (1984) Cryopreservation and fertility of fish, poultry and mammalian spermatozoa. *Proceedings of the 10th Technical Conference on Artificial Insemination and Reproduction,* National Association of Animal Breeders, Columbia, MO, pp. 4–27.
4. Lake, P. E. (1986) The history and future of the cryopreservation of avian germ plasm. *Poult. Sci.* **65,** 1–15.
5. Hammerstedt, R. H. and Graham, J. F. (1992) Cryopreservation of poultry sperm: the enigma of glycerol. *Cryobiology* **29,** 26–40.
6. Bakst, M. R. (1990) Preservation of avian cells, in *Poultry Breeding and Genetics* (Crawford, R. D., ed.), Elsevier, Amsterdam, pp. 91–108.
7. Brock, M. K., Bird, D. M., and Ansah, G. A. (1984) Cryogenic preservation of spermatozoa of the American Kestrel. *Int. Zoo. Yrb.* **23,** 15–20.
8. Sexton, T. J. and Gee, G. F. (1978) A comparative study on the cryogenic preservation of semen from Sandhill Crane and the domestic fowl. *Symp. Zool. Soc. Lond.* **43,** 89–95.
9. Parkes, J. E., Heck, W. R., and Hardaswick, V. (1986) Cryopreservation of peregrine falcon semen and postthaw dialysis to remove glycerol. *Raptor Res.* **20,** 15–20.
10. Samour, J. H., Markham, J. A., Moore, H. D. M., and Watson, P. F. (1988) Semen cryopreservation and artificial insemination in budgerigars *(Melopsittacus undulatus). J. Zool.* **216,** 169–176.
11. Wishart, G. J. (1989) Physiological changes to fowl and turkey spermatozoa during *in vitro* storage. *Br. Poult. Sci.* **30,** 443–452.
12. Tajima, A., Graham, E. D., and Hawkins, D. M. (1989) Estimation of the relative fertilizing ability of frozen chicken spermatozoa using a heterospermic competition method. *J. Reprod. Fert.* **85,** 1–5.
13. Lake, P. E. and Stewart, J. M. (1978) *Artificial Insemination in Poultry.* Ministry of Agriculture, Fish, and Food, Bulletin No. 213. HMSO, London.
14. Ansah, G. A. and Buckland, R. B. (1983) Eight generations of selection for duration of fertility of frozen-thawed semen in the chicken. *Poult. Sci.* **62,** 1529–1539.
15. Bacon, L. D., Salter, D. W., Motta, J. B., Crittenden, L. B., and Ogasawara, F. X. (1986) Cryopreservation of chicken semen of inbred or specialised strains. *Poult. Sci.* **65,** 1965–1971.
16. Lake, P. E. and Stewart, J. M. (1978) Preservation of fowl semen in liquid nitrogen—an improved method. *Br. Poult. Sci.* **19,** 187–194.
17. Lake, P. E., Ravie, O., and McAdam, J. (1981). Preservation of fowl semen in liquid nitrogen: application to breeding programmes. *Br. Poult. Sci.* **22,** 71–77.

18. Buss, E. G. (1993) Cryopreservation of rooster sperm. *Poult. Sci.* **72,** 944–955.
19. Ravie, O. and Lake, P. E. (1984) A comparison of glass ampoules and plastic straws as receptacles for freezing fowl semen. *Cryo-Lett.* **5,** 201–208.
20. Tajima, A., Graham, E. F., and Hawkins, D. M. (1989) Estimation of the relative fertilization ability of frozen chicken spermatozoa using a heterologous competitive method. *J. Reprod. Fert.* **85,** 1–5.
21. Ohara, M., Ozeki, T., Tamura, C., Takahashi, T., and Kusakari, N. (1990) Semen from individual cocks and the fertility of hens after repeated inseminations of frozen semen at 7-day intervals during 4 weeks. *Jpn. Poult. Sci.* **27,** 398–402.
22. Watanabe, M., Terada, T., and Shirakawa, Y. (1977) A diluent for deep freezing preservation of fowl spermatozoa. *J. Fac. Fish. Anim. Husb.* **16,** 59–64.
23. Watanabe, M. and Terada, T. (1980) Studies on the deep freezing of fowl semen in pellet form. *Proc. 9th Intern. Cong. Anim. Reprod. Artif. Insem.* **5,** 477–479.
24. Howarth, B., Jr. (1990) Fertility following intrauterine insemination near the time of oviposition. *Poult. Sci.* **69,** 138–141.
25. Engel, H. N., Froman, D. P., and Kirby, J. D. (1991) An improved procedure for intramagnal insemination of the chicken. *Poult. Sci.* **70,** 1965–1969.

CHAPTER 18

Cryopreservation of Animal and Human Cell Lines

Christopher B. Morris

1. Introduction

The emergence of mammalian cell culture has its origin in the first attempts to culture tissue explants in vitro at the turn of the 20th century *(1–3)*. The subsequent development of complex culture medium formulations in the 1950s *(4,5)* rapidly established a wide range of cell lines and provided a valuable research tool for the study of growth mechanisms and disease. Today, the availability of thousands of animal cell lines offers a reproducible source of material for all aspects of medical and agricultural research.

Maintenance of cell lines in continuous culture by subculture would therefore be impractical because of: cost, i.e., serum containing media is currently about $45/L; risk of exposure to microbial contamination; and the possibilities of culture crosscontamination and genetic drift. Concomitant research into subzero storage methods at the time when cell culture methodology was being developed led to the discovery that addition of glycerol to fowl semen enhanced survival of spermatozoa after storage at –79°C *(6)*. The use of gaseous (> –130°C) and liquid nitrogen (–196°C) now allows indefinite storage of most mammalian cell lines after cryopreservation. The precise mechanism for optimal cryopreservation has been established by studying the effects of freezing and thawing at various rates from which a hypothesis of freezing injury to cells was proposed *(7)*.

The majority of cell lines must be cooled slowly, i.e., −1 to −3°C/min, and thawed rapidly to achieve maximum viability. The rate of cooling is optimized to allow time for intracellular water to escape, and subsequently reduce the amount of intracellular ice formed. The presence of intracellular ice during thawing, i.e., postfreezing resuscitation, may cause lethal damage to intracellular membranes *(8)*. Addition of cryoprotectant to the cells depresses the temperature at which intracellular ice is formed and allows cooling rates to be reduced for more efficient water loss *(9)*.

The development of program-controlled freezers now allows individual cooling profiles to be designed that give maximum cell viability during cryopreservation, i.e., identical pre- and postfreezing viabilities. Within culture collections such as the European Collection of Animal Cell Cultures (ECACC; Salisbury, UK), many thousands of cell lines are now currently available to which new cell lines are added annually. A systematic approach to the quality control of cell lines and their cell banks is therefore essential to ensure future supplies of authentic material *(10,11)*.

2. Materials

1. An appropriate cabinet located in the correct containment level must be used in a class II microbiological safety cabinet in a containment level 2 laboratory, or a class III cabinet in a containment level 3 laboratory *(12)*. The handling requirements of any cell line should be assessed using the guidelines issued by national regulatory bodies in consultation with your safety officer, prior to their introduction into the laboratory *(13)*.
2. Cell cultures: These should be in active growth, i.e., log-phase, usually 2–4 d after subculture. Cells that have entered stationary phase should not be used (*see* Note 1).
3. Freeze medium: This will either be the growth medium supplemented with 20–25% (v/v) serum and 10% (v/v) cryoprotectant or whole serum and 10% (v/v) cryoprotectant. The choice of cryoprotectant will be determined by the cell type, but for the majority of cell lines dimethyl sulfoxide (DMSO) or glycerol can be used. An alternative is polyvinyl pyrrolidine, a high mol wt polymer (*see* Note 2).
4. Cryovials or ampules: These are obtained from tissue-culture plasticware supplies, e.g., Nunc (Roskilde, Denmark), Becton-Dickinson (Bedford, MA), Corning (Corning, NY), as presterilized (irradiated) screwcap 1.0- or 1.8-mL vials (*see* Note 3). Some manufacturers also supply special racks for holding the vials during filling.

5. Freezing system: The best and most reproducible method is the programmable freezer, e.g., Planer Products (Sunbury-on-Thames, UK), and Sy-Lab (Austria) (*see* Notes 4 and 5).
6. Storage system: Liquid nitrogen storage vessels with inventory systems suitable for cryovials. Most vessels can be arranged to store vials in gas, liquid, or a combination of both. Automatic filling (top-up) and alarm systems are advisable to prevent accidental loss of stored material (*see* Note 6).
7. Improved Neubauer hemocytometer and 0.4% (w/v) trypan blue in phosphate-buffered saline (PBS) for calculating both total and viable cell numbers.
8. Small liquid nitrogen vessel for transporting ampules.
9. Protective full face mask, cryogenic (insulated) gloves, waterproof apron, long forceps, and clamping scissors.

3. Method

1. Cell lines must be handled in appropriate laboratory conditions providing adequate protection to the operator. Only one cell line should be handled at a time, with separate reagents used for each cell line to avoid the risks of microbial and cross–contamination.
2. Microscopically examine all cultures to be frozen for cell morphology, density, and microbial contamination using a good quality inverted phase microscope fitted with at least a x10 and x20 objective (i.e., x100 and x200 final magnification). Reject any suspect cultures.

    Normally, the cell density should not exceed 85% of its maximum growth density and they should have been passaged at least twice in the absence of all antibiotics prior to freezing (*see* Note 7).
3. Count the cells after staining with trypan blue to estimate the percentage of viable cells in the culture. Suspension cells can be counted directly by diluting 100 µL between 2-fold (1:1) and 10-fold (1:9) with the trypan blue.

    Adherent cells will require a proteolytic enzyme, e.g., trypsin or trypsin + ethylenediaminetetraacetic acid (EDTA) to remove the cell sheet. Cells should be prepared as for routine subculture, remembering to neutralize the enzyme by addition of serum containing medium or soya bean inhibitor in the case of serum free cultures. Dilute an aliquot of cells in trypan blue.
4. Load a prepared hemocytometer with the diluted cells using a file tip Pasteur or micropipet. Allow the mixture to be drawn under the coverslip by capillary, rather than active pipeting. Fill the chamber completely.
5. Count the cells over one of the nine 1 mm^2 squares (bright, retractile cells are viable and dark blue cells are dead) using a phase microscope (an inverted type is suitable). Repeat the process over three more squares. Usu-

ally the corner squares are used. For statistically accurate counts, a range of 30–100 cells/mm^2 should be counted. Prepare another sample if the counts are outside this range.
6. Estimate the total and viable cell count as follows:

    cells/mL = (No. of cells counted/No. of 1 mm^2 squares counted) × dilution × 10^4

    The percentage of viable cells is:

    [Total viable cells/Total cell count (viable + dead)] × 100

    Healthy cultures should exceed 90% viability. Low viabilities or the presence of large quantities of cell debris are an indication of suboptimal culture conditions or exhaustion of the nutrient supply.
7. Calculate the volume of cells required to fill the ampules, e.g., 10 ampules at 5 × 10^6 cells/ampule = 50 × 10^6 cells. A recommended number of cells/ampule is between 4–10 × 10^6 cells.
8. Centrifuge the cells, using the minimum g force necessary to sediment them, e.g., 100g for 5 min.
9. Decant the medium and resuspend the cells in the freeze medium to the required cell density (*see* Note 8). To aid resuspension, vibrate the pellet gently with a finger after decanting the medium.
10. Dispense the cells into premarked ampules in 1-mL aliquots. Ampules must be clearly marked with the cell designation, passage number, freeze batch number, and date of freezing.
11. Keeping the ampules vertical, to avoid spillage into the cap, transfer them to the freezer (*see* Section 2., item 5 and Notes 4 and 5).
12. Once frozen, transfer the ampules to nitrogen storage. Wear a face mask and full protective clothing. Record the ampule location. Graphical databases specifically designed for use with cryogenic systems are available from I/O Systems Ltd. (Ashford, UK).
13. Check at least one ampule from each batch, frozen (cell bank) for viability and growth potential. Always allow at least 24 h storage in their final location before starting quality control tests.
14. Transfer an ampule to the top of the storage vessel using long forceps or clamping scissors, placing it in an aluminum screwtop canister. Full protective clothing must be worn when removing ampules from storage.
15. Transport the ampule to a water bath, after temperature equilibration, either in a small dewar with nitrogen or preferably in dry-ice. It is essential not to fully reimmerse ampules in liquid nitrogen. Therefore, if a dewar is used, place ampules through holes in a piece of polystyrene that will float on the liquid surface, keeping the screwthread above liquid.

16. Thaw ampules in a water bath set at the cells normal growth temperature, 37°C for mammalian cells, 25°C for amphibian cells. Rapid and complete thawing is vital to retain viability. Again, float ampules in racks or polystyrene, do not submerge.
17. Transfer ampules to a microbiological safety cabinet and wipe the ampule surface with 70% (v/v) ethanol. Remove the contents using a sterile Pasteur or 1-mL pipet and transfer to a 15-mL screwcap centrifuge tube. Add dropwise 2 mL of antibiotic–free growth medium, mixing gently by swirling. Another 2 mL of medium is added at normal speed. Remove 100 µL for total and viable cell counts (*see* steps 3–6). Set up new cultures at between 30 and 50% of their maximum cell density (*see* Note 9).
18. Maintain the culture for at least 5 d to monitor cell growth and absence of microbial contamination. Master cell banks should be fully quality controlled to ensure their authenticity (*see* Note 10).

4. Notes

1. Cells harvested for cryopreservation should be at their optimum viability to ensure maximum survival during freezing and after thawing. This is especially relevant when the method of cryopreservation reduces the number of viable cells and increases the chances of selecting freeze-tolerant populations that may have different characteristics from the original population.
2. DMSO is sterilized by filtration using 0.2-µm DMSO-resistant filters. Glycerol can be sterilized by autoclaving. DMSO is toxic if left in contact with cells for more than a short period of time. Once the cells have been prepared for freezing, they should be ampuled and frozen as soon as possible. Addition of DMSO to culture medium normally increases the pH, especially if it contains sodium bicarbonate. This increase will further reduce cell viability at values above pH 8. It may then be necessary to gas the medium with 5–10% (v/v) CO_2 in air. An alternative freeze medium is whole serum, i.e., fetal bovine serum (FBS) or newborn calf serum (NBCS), to which the cryoprotectant is added. This has the double advantage of greater pH control, and protection against freeze damage owing to the increased levels of albumins.

    A further consideration, often overlooked, is the country of origin of the serum used to grow and freeze the cells. If cells are to be used in commercial processes or shipped to other countries, e.g., the US, they must be shown to be free of contaminants and adventitious agents before an import permit is issued. Using serum from countries of origin other than those designated Zone 1 may lead to difficulties in importation. Consult your supplier for the current situation.

When cells have been grown in serum-free medium it may be desirable to omit serum from the freeze medium. This can reduce the viability of the cells during freezing, because of the protective nature of serum on surface membrane components. Addition of 0.1% (w/v) methyl cellulose in the freeze medium has been found to reduce cell death *(14)*.

Cell lines requiring complex media or addition of growth factors should initially be frozen with these additions included in the freeze medium, e.g., Interleukin 2 for the mouse T-cell cytotoxic T-lymphocyte (CTLL). Their inclusion may help to stabilize surface proteins acting as receptors for the growth factors.

If zwitteronic buffers (e.g., HEPES, Tricine) are usually included in the growth medium, they must be excluded from the freeze medium to avoid hypertonic stress during cooling. Antibiotics should never be included.

3. Cryovials can be obtained with either an internal or external thread. In long-term storage, both types will allow entry of nitrogen liquid or gas. On removal from storage, extreme caution must be exercised to prevent explosion of the cryovial because of sudden expansion of the trapped nitrogen. Cryovials that have been stored in liquid should be allowed to equilibrate to the temperature at the top of the storage vessel, i.e., gas phase, before transferring to a water bath or laboratory. It is advisable to place ampules during this period in a screw-top aluminum canister to contain any exploded material for decontamination *(see* Note 5).
4. To retain maximum viability during cryopreservation, cells must be cooled at a constant slow rate, -1 to $-5°C/min$ *(15)*. Programmable freezers are therefore the only means of achieving total viability, i.e., recovery of the same number of viable cells after freezing, as before freezing. This is because they can increase the cooling rate at the most critical point in the program, the eutectic point, when the cells freeze and release energy owing to the latent heat of fusion, usually between -4 and $-8°C$. If rapid cooling does not occur at this point to compensate for the increase in temperature, the cells warm up with subsequent cell injury.
5. A less expensive alternative is the two-stage freezer. Ampules are placed in the neck of a nitrogen dewar that contains a low level of liquid nitrogen, exposing them to the gaseous phase at a point where the temperature is about -25 to $-30°C$. After 20–30 min the ampules should have frozen and are then plunged into the liquid, prior to transfer to their final storage location.

The least expensive and reliable method is to place the ampules in a heavily insulated box, e.g., polystyrene, either directly into precut holes or wrapped in a paper towel and place the box at $-80°C$ for 24 h. The ampules are then transferred to their final storage location. A better alternative to polystyrene is a cooling box produced by Nalgene (Cat No 5100/001) which cools at −1°C/min when placed at −80°C.

6. To safeguard against loss of cryopreserved material caused by sudden vacuum loss or staff forgetting to fill vessels, it is essential to install auto-fill and alarm systems on storage vessels. Many large volume vessels already include these facilities. Storage vessels can be automatically supplied from 100- to 200-L self-pressurizing reservoirs, placed in the same room. Larger reservoirs will require special housing arrangements. Local alarms provide audible warning of problems, but to fully cover all emergencies they need to be connected to a telemetric system that will signal a radiopager. An additional precaution is to divide up the cell banks and store them in several vessels. It is essential that storage vessels are located in a ventilated room, which will be checked regularly. When large numbers of vessels or large volumes of nitrogen are used, an oxygen monitor must be filled in the room, i.e., under the guidance of your safety officer and nitrogen supplier. At ECACC the monitor has been connected to an automatic ventilation system that operates when the oxygen level reaches 18.5% (v/v).
7. Routine addition of antibiotics to cell cultures must be avoided at all times, especially in stock cultures. A properly equipped and organized laboratory can avoid this problem with good technique. As most antibiotics will only suppress persistent infections, which in their absence could be easily identified and eliminated, it is essential that cells to be cryopreserved are known to be free of infection.
8. If whole serum with cryoprotectant is used, this can be prepared in advance and checked for microbial contamination by removing a 5–10% sample and incubating with thioglycollate and tryptone soya broths for 7 d. Aliquot the freeze medium into appropriate volumes and store at −20°C or below. However, preparation of freeze media containing cell culture medium (e.g., MEM, RPMI 1640) is necessary just prior to use to avoid pH changes on standing. To minimize contamination risk, all components should be pretested. Resuspending cells for freezing in precooled freeze medium, 0–4°C, may improve their survival prior to freezing. When this approach is used, it is important to maintain the cells at a constant temperature during all subsequent handling by placing ampules in ice until they are frozen.
9. After thawing cells it is necessary to slowly dilute the cryoprotectant to prevent osmotic shock. The need to remove the cryoprotectant will depend on how much the cells are diluted (above 10 mL is usually sufficient to overcome toxic effects). DMSO will also evaporate from the medium at 37°C.

When it is necessary to centrifuge the cells, use the minimum g force to sediment them to prevent shearing damage, i.e., $70-100g$. To initiate rapid growth, it is advisable to inoculate new cultures at a higher density than for routine subculture, e.g., between 3 and 4×10^5 viable cells/mL for most suspension cells and between 3 and 4×10^4 viable cells/cm^2 for adherent cells. Monitor their growth and subculture once they reach a maximum density. At the ECACC, cell banks are frozen in a programmable freezer. The majority of postthaw viabilities exceed 85% and are normally only a few percent lower than the prefreezing viability. Those with postthaw viabilities below 75% are rejected. Viabilities of cells frozen by other methods tend to be much lower, and consequently resuscitated cultures contain a high level of debris and dead cells, which may have an inhibitory effect on the remaining viable cells. Debris can be removed from adherent cells by allowing the culture to grow for 24 h and then changing the medium (washing the attached cells with medium first, if necessary). Suspended cells present a greater problem, and may require a sedimentation stage to separate viable and dead cells.

10. The minimum number of tests that should be carried out on master cell banks are total and viable cell counts, growth potential, and screening for bacteria and fungi. Quality control and the principles of cell banking should be an established part of the laboratory procedures if you are to establish problem-free cell stocks. Other tests that therefore became essential to comply with these criteria are:

   a. Screening for mycoplasma: These are very small microorganisms in a size range below most bacteria. Therefore, although they may be present at concentrations between 10^6 and 10^8 organisms/mL, they do not usually cause turbidity in cultures and remain undetected. Mycoplasma constitute one of the greatest problems in cell culture and cultures should be routinely checked for their presence *(16)*.

   b. Cell line authenticity: Cross-contamination of cell cultures has been, and still is to some extent, a major problem *(17)*. The species of origin of a cell line can be identified using isoenzyme analysis *(18)*. A kit is now available that is suitable for use in any laboratory (Authentikit, Innovative Chemistry, Inc., PO Box 90, Marshfield, MA 02050). Cytogenetic analysis is used to identify normal and abnormal karyotypes as well as species but requires considerable expertise to interpret *(19)*. More recent techniques are DNA fingerprinting *(20)* and gel analysis of restriction digests, both of which are used at ECACC for species verification of cell lines and master cell banks.

References

1. Harrison, R. G. (1907) Observations on the living developing nerve fibre. *Proc. Soc. Exp. Biol. Med.* **4,** 140–143.

2. Carrel, A. (1912) On the permanent life of tissues outside the organism. *J. Exp. Med.* **15**, 516–528.
3. Rous, P. and Jones, F. S. (1916) A method for obtaining suspensions of living cells from the fixed tissues, and for the plating out of individual cells. *J. Exp. Med.* **23**, 549–555.
4. Morgan, J. F. (1950) Nutrition of animal cells in tissue culture. Initial studies on a synthetic medium. *Proc. Soc. Exp. Biol. Med.* **73**, 1–8.
5. Healy, G. M., Fisher, D. C., and Parker, R. C. (1954) Nutrition of animal cells in tissue culture IX synthetic medium No 703. *Can. J. Biochem. Physiol.* **32**, 327–337.
6. Polge, C., Smith, A. U., and Parkes, A. S. (1949) Revival of spermatozoa after vitrification and dehydration at low temperatures. *Nature* **164**, 666.
7. Lovelock, J. E. (1953) The mechanism of the protective action of glycerol against hemolysis by freezing and thawing. *Biochim. Biophys. Acta* **11**, 28–36.
8. Mazur, P. (1977) The role of intracellular freezing in the death of cells cooled at supraoptimal rates. *Cryobiology* **14**, 251–272.
9. Diller, K. R. (1979) Intracellular freezing of glycerolized red cells. *Cryobiology* **16**, 125–131.
10. Doyle, A. and Morris, C. B. (1991) Maintenance of animal cells, in *Maintenance of Microorganisms and Cultured Cells* (Kirsop, B. E. and Doyle, A., eds.), Academic, London, pp. 227–242.
11. Bolton, B. J., Morris, C. B., and Mowles, J. M. (1993) General principles of cell culture, in *Methods of Immunological Analysis,* vol. 3 (Masseyeff, R. F., Albert, W. H., and Staines, N. A., eds.), Verlag Chemie.
12. HMSO Advisory Committee on Dangerous Pathogens (1990) *Categorisation of Pathogens According to Hazard and Categories of Containment,* Author, London.
13. Frommer, W., Archer, L., Boon, B., Brunius, G., Collins, C. H., Crooy, P., Doblhoff-Dier, O., Donikian, R., and Economidis, J. (1993) Safe biotechnology (5). Recommendations for safe work with animal and human cell cultures concerning potential human pathogens. *Appl. Microbiol. Biotechnol.* **39**, 141–147.
14. Ohno, T., Kurita, K., Abe, S., Eimori, N., and Ikawa, Y. (1988) A simple freezing medium for serum-free cultured cells. *Cytotechnology* **1**, 257.
15. Grout, B., Morris, J., and McLellan, M. (1990) Cryopreservation and the maintenance of cell lines. *Trends Biotechnol.* **8**, 293–297.
16. Mowles, J. M. (1990) Mycoplasma detection, in *Methods in Molecular Biology,* vol. 5, *Animal Cell Culture* (Pollard, J. W. and Walker, J. M., eds.), Humana, Clifton, NJ, pp. 65–74.
17. Nelson-Rees, W. A., Daniels, D. W., and Flandermeyer, R. R. (1981) Cross contamination of cells in culture. *Science* **212**, 446–452.
18. Halton, D., Peterson, W., Jr., and Hukku, B. (1983) Cell culture quality control by rapid isoenzymatic characterisation. *In Vitro* **19**, 16–24.
19. Macgregor, H. and Varley, J. (1988) *Working with Animal Chromosomes.* Wiley, Chichester, UK.
20. Stacey, G. N., Bolton, B. J., and Doyle, A. (1991) The quality control of cell banks using DNA fingerprinting, in *DNA Fingerprinting: Approaches and Applications* (Burke, T., Dolt, G., Jeffreys, A., and Wolff, R., eds.), Birkhauser Verlag, Basel, Switzerland.

CHAPTER 19

Cryopreservation of Semen from Domestic Livestock

Mark R. Curry

1. Introduction

Although the first observations concerning low temperature preservation of spermatozoa date back to 1776 (when Spallanzani noted that spermatozoa, cooled in snow, became inactive but were revived on warming), successful cryopreservation protocols truly date from only the 1940s and 1950s. Around this time semen preservation was in the forefront of the new science of cryobiology and was important in the formulation of the first theories concerning the causes of cryoinjury. In 1938 Jahnel noted the survival of human spermatozoa stored at the temperature of solid carbon dioxide, an observation taken up by Parkes in 1945 *(1)*, but the single most important development came with the discovery by Polge, Smith, and Parkes in 1949 *(2)* that glycerol could act as a cryoprotectant for spermatozoa. The initial experiments with glycerol were performed using fowl spermatozoa but were quickly followed by the successful preservation of bull spermatozoa *(3)* with the first calf from artificial insemination (AI) using frozen/thawed spermatozoa reported in 1951 *(4)*. However, early success with the bull rather belied the difficulties that became apparent with other domestic species. The use of glycerol had allowed bull spermatozoa to survive at the comparatively slow freezing rates obtained using methanol cooled with solid CO_2 and with semen contained in large glass-freezing ampules. When other domestic species were tried, the levels of glycerol tolerated by bull spermatozoa proved to be damaging. The problems encountered were compounded by difficulties in assessing the success of cryopreservation protocols, as morphologically intact and apparently normally motile cells proved on insemination to be functionally damaged and consequently infertile. Technical advances, with the replacement of solid carbon dioxide by liquid nitrogen for freezing and long-term storage, brought faster freezing rates obtained either by using small volume plastic straws suspended in liquid nitrogen vapor *(5)* or by the pellet freezing method (small drops of semen on to the surface of solid CO_2) *(6)*. Faster freezing rates permitted the use of lower concentrations of glycerol, leading to some success with frozen ram *(7)* and boar semen *(8)*.

The great variation between species in sperm cryosurvival results from differences in size and shape of spermatozoa and in the properties of their membranes. There also exists variation between species in the fertility of frozen semen used for AI, although this may be explained in part by differences in the number of spermatozoa used, in the site of deposition, and in the degree of technical difficulty involved in the insemination procedure. Attempts to overcome this species variation have produced an extensive literature detailing the largely empirical optimization of freezing protocols that has taken place over the last 30 yr. The effects of varying diluent composition, freezing and thawing rates, and freezing methodology together with the various interactions that exist between these factors have been reported for a wide range of species *(9,10)*.

In 1866 Mantegazza recorded the survival of human spermatozoa frozen at $-17°C$ and subsequently wrote:

> If the human sperm can be preserved for more than four days at the temperature of melting-ice without undergoing any change, then it is certain that scientists of the future will be able to improve breeds of horses and cattle without having to spend enormous sums of money in transporting thoroughbred stallions and bulls. It will be possible to carry out artificial insemination with frozen sperm sent rapidly from one locality to another. It should also be feasible for a husband who dies on the battlefield to fertilise his wife and thus to have legitimate sons even after his own death.

This remarkable prediction has been realized in full 125 yr later. However, although offspring for all the major domestic species have been obtained using frozen/thawed semen, it is still only with bull spermatozoa that freezing is used on a large commercial scale, and that comparable fertility is obtained using fresh and frozen semen. In all other species there is a considerable reduction in fertility that must be countered by inseminating with greater sperm numbers. Obviously much remains to be done if we are to exploit fully the great opportunities offered by semen cryopreservation.

The protocol given in this chapter, although based on that widely used for bull spermatozoa, is not given as being optimal for any particular species but as one that should result in a degree of survival over a range of species. The Notes section gives some further information concerning the specific requirements of different species.

2. Materials

1. TEST buffer: *N*-tris[Hydroxymethyl]methyl-2-aminoethane-sulfonic acid (TES) 48.6 g/L, Tris[Hydroxymethyl]amino-methane (TRIS) 11.0 g/L. Dissolve TES and TRIS in twice distilled water and adjust pH to 7.1. Filter through a 0.2-μm filter and store at 4°C. Although TEST buffer may be stored at 4°C prior to use, the complete TEST-egg yolk-glycerol (TEST-EY-G) freezing diluent is best prepared fresh on the day of use. All reagents should be Analar grade.
2. TEST-EY-G: To TEST buffer solution add: 9 g/L glucose, 500 IU/mL penicillin, 0.5 mg/mL streptomycin sulfate. Adjust osmolarity to 350 mOsm. Add 15% (v/v) egg yolk. Egg yolk must be completely separated from the albumin and dried by rolling on filter paper. Centrifuge at 2000*g* for 20 min and decant supernatant for use discarding the pellet, then add 3% (v/v) glycerol.
3. 0.25 mL Cryostraws and polyvinyl acetate (PVA) powder.
4. Dewar containing liquid nitrogen suitable for long-term storage of the frozen samples.
5. Safety equipment: cryogloves, goggles, face-mask, and so on.

3. Method

1. Dilution: Assess the sample for concentration and motility (*see* Note 1). Dilute in TEST-EY-G to a final concentration of approx 500×10^6/mL; if this involves a dilution of less than 5:1 TEST-EY-G to semen, wash the semen once in TEST-EY-G and resuspend to the final concentration in TEST-EY-G (*see* Notes 2–8). Load the diluted semen into 0.25-mL plastic straws and seal with PVA powder (*see* Note 9).
2. Cooling to 5°C: Place the straws in a 500-mL beaker of water at ambient temperature and place in a refrigerator for 3 h to allow slow cooling to 5°C.
3. Equilibration period: The total time elapsed between semen collection and commencement of the freezing process should not be less than 5 h. A portion of this period will be taken up with the processing and slow cooling stages for the remainder of the time; hold the straws in the refrigerator at 5°C (*see* Notes 10 and 11).

4. Freezing: Suspend the straws horizontally 5 cm above the surface of the liquid nitrogen for 7 min and then plunge rapidly into the liquid nitrogen (*see* Note 12).
5. Long-term storage: Keep the straws submerged in liquid nitrogen—in this state they may be stored indefinitely with no further deterioration. Straws allowed to partially thaw before being resubmerged will suffer some degree of sperm damage.
6. Thawing: Remove straws from the liquid nitrogen and thaw in air for 3 s, then plunge into a water bath at 35°C for 15 s (*see* Notes 13 and 14).

4. Notes

1. Semen assessment: Semen of poor quality is generally found to have poor survival on freezing and thawing. A routine assessment should be performed to ensure samples conform to minimum standards for concentration, motility, and normal morphology.
2. Diluent: A wide range of diluents of varying composition have been used in successful freezing protocols, but the basic requirements for a freezing diluent are the same in all cases. The diluent must maintain osmolarity, pH, and ionic strength; it must provide an energy substrate, contain some cryoprotective agent, and, in the majority of cases, contain antimicrobial agents. The majority of diluents are hyperosmotic relative to semen, although isotonic diluents have been successfully used. Hypertonic diluents probably result in a degree of dehydration that is advantageous during freezing.
3. The optimum pH for mammalian spermatozoa is close to neutrality, most used diluents buffer to pH 6.9–7.1. TES-TRIS buffer, as in TEST-EY-G, acts as a major component in the diluent serving to control pH and tonicity. Citrate, one of the earliest and most extensively used buffers which also acts as a major component of other buffers (e.g., MES or HEPES), may be used as minor components controlling pH only. Proteins present in either fresh- or skim-dried milk diluents are adequate to buffer pH and milk diluents have been used extensively for a range of species *(11,12)*. Egg yolk proteins also have a minor buffering effect.
4. Ionic strength: Ionic strength appears to be of minor importance; sperm have been successfully frozen at widely differing ionic strengths provided osmolarity is maintained by nonionic constituents. Some species (e.g., boar) do show optimal survival at relatively low ionic strengths *(13)*.
5. Energy substrate: Energy substrate is provided by glycolysable sugars, such as glucose, fructose, or mannose, species differences in the rate at which these sugars are metabolized may govern which is optimal for any given species. Sugars also have other functions within the diluent contrib-

uting to the osmotic strength and in some cases having a cryoprotective effect *(14)*. However, the literature is confusing in that different combinations of sugars have been found to be beneficial under different conditions.

6. Cryoprotective agents may be divided into two major classes, nonpenetrating and penetrating. Glycerol, the almost universal cryoprotectant for spermatozoa, is an example of a penetrating cryoprotectant, although it may also function as a nonpenetrating protectant *(15)*. The cryoprotective effects of any sugars included in the diluent are an example of a non-penetrating cryoprotectant. The concentration of glycerol in the diluent for optimal cryosurvival varies for different species and in some species high concentrations may preserve motility but impair fertility. For example, boar spermatozoa show good postthaw motility when frozen with 7% (v/v) glycerol but fertility is only maintained with less than 3% (v/v) *(16)*. Optimal glycerol concentrations for other domestic species are: bull, 6–9% (v/v) *(17)*; ram, 3–4% (v/v) *(7)*; and stallion, 4% (v/v) *(18)*. In some cases it has been suggested that the adverse effects of glycerol may be minimized by adding it to the semen at 5°C. A two-step dilution process for the addition of glycerol has also been used with bull semen to minimize both toxic and osmotic damage, but other studies have shown little advantage and any benefit of these additional precautions remains unproven. One exception to the use of glycerol is for rabbit spermatozoa where dimethyl sulfoxide (DMSO) is the cryoprotectant of choice *(19)*.

   Egg yolk also has a cryoprotective role in the diluent acting to protect cells against cold shock on cooling below ambient temperature and during freezing. The mechanism of this protection is not fully understood but has been isolated to the low density lipoprotein fraction of the egg yolk and may depend on egg yolk providing a source of lipid for incorporation into membranes *(20)*. As with glycerol, the optimal egg yolk level varies between different species.

7. Antimicrobial agents: Seminal fluid is normally biologically sterile, however, contamination is generally unavoidable during semen collection. The presence of microorganisms can have a range of adverse effects on fertility and the inclusion of antimicrobial agents has been variably reported as having a beneficial effect on the outcome of insemination, probably reflecting variation in initial contamination levels. A broad spectrum antibiotic as a safeguard against microorganism proliferation is included in most diluents. Penicillin (500–1000 IU/mL) and streptomycin (0.5–1.0 mg/mL) are the most commonly used and at these levels are nontoxic to spermatozoa *(21)*, a number of other antibiotics have been used, although some have been found to have adverse effects on sperm metabolism, e.g., sulfanilamide, oxytetracycline, and gentamycin *(22,23)*.

8. The dilution stage has three purposes: to minimize any potential toxic effects of seminal plasma, to suspend the spermatozoa in the TEST-EY-G freezing diluent, and to extend the semen to allow maximum usage. Overdilution may lead to adverse so-called dilution effects whereas with very low dilution rates, care must be taken that the spermatozoa are exposed to the correct levels of egg yolk and glycerol in the freezing diluent. Dilution rate is usually governed in practice by the number of spermatozoa required for AI. The relatively high postthaw fertility of bull spermatozoa means comparatively high dilution rates are possible as a smaller number of spermatozoa are required in the inseminate (approx 2×10^7) *(24)*, thus maximizing the number of inseminations per ejaculate; much greater sperm numbers are required for insemination with the boar (approx 8×10^9) *(13)*, the ram (approx 3×10^8) *(7)*, and the stallion (approx 2.5×10^8) *(25)*, reflecting the lower fertility of the frozen/thawed semen and consequently the lower dilution rates in these species.
9. Plastic straws are available in a range of sizes from two major suppliers Instruments de Medicine Veterinaire (IMV; l'Aigle, France) and Minitüb Tiefeubach bei Laudshut, Germany). Different colored straws and seals enable samples to be color-coded for ease of identification. Automated systems are available for filling, sealing, and labeling large numbers of straws.
10. Slow cooling to 5°C is required to minimize the cold shock damage apparent when spermatozoa are cooled below 15°C. Sensitivity to cold shock varies between species. Stallion and human sperm are fairly cold shock resistant, whereas bull, ram, and particularly boar spermatozoa are cold shock sensitive *(26)*. The presence of egg yolk in the diluent serves to protect cells against cold shock *(20)*.
11. Sperm cryosurvival is generally improved if the cells are held at ambient temperature or at 5°C for a period of time before freezing. This holding period was originally believed to allow glycerol penetration, however, glycerol is now known to permeate quite rapidly and the equilibration period is thought to involve a process of membrane stabilization. Membrane stabilization confers cold shock resistance. Boar sperm in particular require several hours to become resistant to cold shock *(26)*. Acquired resistance to cooling above 0°C may confer additional resistance to freezing and thawing.
12. Optimal cooling rates for semen of domestic species are generally considered to be in the range of 10–100°C/min. The protocol given here gives a rate of approx 90°C/min, but this is not a linear cooling rate. Spontaneous ice nucleation occurs at approx −15 to −20°C, and the latent heat released causes a temperature rise of around 7 or 8°C. Thereafter, the cooling curve is sigmoidal (Fig. 1). The final temperature reached before plunging into the liquid nitrogen is approx −120°C. Actual cooling rates are dependent

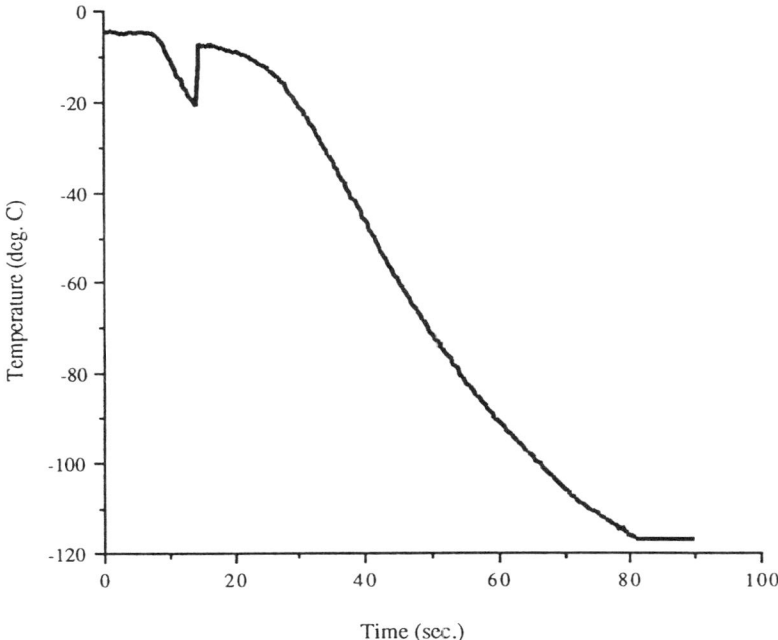

Fig. 1. Graph showing freezing rate for 0.25-mL straws suspended 5 cm above liquid nitrogen surface. (This graph is obtained using a 20-cm diameter dewar with a stable vapor column 24 cm in depth. The dimensions of the system used and the number of straws loaded will effect the freezing rate obtained.)

on the conditions of the vapor column and the number of straws frozen. Slower cooling rates may be achieved by suspending the straws further from the liquid nitrogen surface. However, the straws should be cooled to at least −50°C before plunging.
13. Faster thawing rates may be achieved by using higher temperatures for shorter periods of time, e.g., 65°C for 5 s. Faster thawing is generally considered to be the most effective, although the evidence for enhanced fertility is not good. At faster thawing rates care must be taken not to allow any overheating of the sperm.
14. A number of up-to-date reviews exist that give an introduction to the very extensive literature concerning semen cryopreservation. Reviews are available concerning semen preservation and AI in general *(9,10)*, recent cryotheory with respect to cryopreservation of spermatozoa *(27)*, and for the major domestic species bull *(24)*, boar *(28)*, ram *(16)*, and stallion *(29)*.

References

1. Parkes, A. S. (1945) Preservation of human spermatozoa at low temperatures. *Br. Med. J.* **2,** 212,213.
2. Polge, C., Smith, A. U., and Parkes, A. S. (1949) Revival of spermatozoa after vitrification and dehydration at low temperatures. *Nature* **164,** 666.
3. Smith, A. U. and Polge, C. (1950) Survival of spermatozoa at low temperatures. *Nature* **166,** 668.
4. Stewart, D. (1951) Storage of bull spermatozoa at low temperatures. *Vet. Rec.* **63,** 65,66.
5. Cassou, R. (1964) La méthode des paillettes en plastique adaptée à la généralisation de la congélation. *V International Congress on Animal Reproduction and Artificial Insemination*, Trento, Italy, pp. 540–546.
6. Nagase, H. and Niwa, T. (1964) Deep freezing of bull semen in concentrated pellet form. 1. Factors affecting the survival of spermatozoa. *V International Congress on Animal Reproduction and Artificial Insemination*, Trento, Italy, pp. 410–415.
7. Colas, G. (1975) Effect of initial freezing temperature, addition of glycerol and dilution on the survival and fertilizing ability of deep-frozen ram semen. *J. Reprod. Fertil.* **42,** 277–285.
8. Wilmot, I. and Polge, C. (1977) The low temperature preservation of boar spermatozoa. 3. The fertilizing capacity of frozen and thawed boar semen. *Cryobiology* **14,** 483–491.
9. Watson, P. F. (1979) The preservation of semen in mammals, in *Oxford Reviews of Reproductive Biology* (Finn, C. A., ed.), Oxford University Press, Oxford, UK, pp. 283–350.
10. Watson, P. F. (1990) Artificial insemination and the preservation of semen, in *Marshall's Physiology of Reproduction,* vol. 2, 4th ed. (Lamming, G. E., ed.), Churchill Livingstone, London, pp. 747–869.
11. Fiser, P. S., Ainsworth, L., and Langford, G. (1981) Effect of osmolarity of skim-milk diluent and thawing rate on cryosurvival of ram spermatozoa. *Cryobiology* **18,** 399–403.
12. Melrose, D. R. (1962) Artificial insemination in cattle, in *The Semen of Animals and Artificial Insemination* (Maule, J. P., ed.), Commonwealth Agricultural Bureaux, Farnham Royal, England, pp. 1–181.
13. Paquignon, M. (1984) Semen technology in the pig, in *The Male in Farm Animal Reproduction* (Courot, M., ed.), Martinus Nijhoff, Boston, MA, pp. 202–218.
14. Nagase, H., Yamashita, S., and Irie, S. (1968) Protective effects of sugars against freezing injury to bull spermatozoa. *VI International Congress on Animal Reproduction and Artificial Insemination*, Paris, France, pp. 1111–1113.
15. Berndtson, W. E. and Foote, R. H. (1972) The freezability of spermatozoa after minimal pre-freeze exposure to glycerol or lactose. *Cryobiology* **9,** 57–60.
16. Graham, E. F., Crabo, B. G., and Pace, M. M. (1978) Current status of semen preservation in the ram, boar and stallion. *J. Animal Sci.* **47(Suppl. II),** 80–118.
17. Mortimer, R. G., Berndtson, W. E., Ennen, B. D., and Pickett, B. W. (1976) Influence of glycerol concentration and freezing rate on post-thaw motility of bovine spermatozoa in continental straws. *J. Dairy Sci.* **59,** 2134–2137.

18. Cochran, J. D., Amann, R. P., Froman, D. P., and Pickett, B. W. (1984) Effects of centrifugation, glycerol level, cooling to 5°C, freezing rate and thawing rate on the post-thaw motility of equine sperm. *Theriogenology* **22,** 25–38.
19. Wales, R. G. and O'Shea, T. (1968) The deep-freezing of rabbit spermatozoa. *Aust. J. Biol. Sci.* **21,** 831–833.
20. Watson, P. F. (1981) The roles of lipid and protein in the protection of ram spermatozoa at 5°C by egg-yolk lipoprotein. *J. Reprod. Fertil.* **62,** 483–492.
21. Dunn, H. O., Bratton, R. W., and Henderson, C. R. (1953) Fertility of bovine spermatozoa in buffered whole egg extenders containing penicillin, streptomycin, sulphonamide and added glucose. *J. Dairy Sci.* **36,** 524–529.
22. Berndtson, W. E. and Foote, R. H. (1976) Survival and fertility of antibiotic treated bovine spermatozoa. *J. Dairy Sci.* **59,** 2130–2133.
23. Ahmad, K. and Foote, R. H. (1984) Survival of frozen bull spermatozoa in extenders treated with different antibiotics. *J. Dairy Sci.* **67,** 152.
24. Pickett, B. W. and Berndtson, W. E. (1974) Preservation of bovine spermatozoa by freezing in straws: a review. *J. Dairy Sci.* **57,** 1287–1301.
25. Cristanelli, M. J., Squires, E. L., Amann, R. P., and Pickett, B. W. (1984) Fertility of stallion semen processed, frozen and thawed by a new procedure. *Theriogenology* **22,** 39–45.
26. Watson, P. F. (1981) The effects of cold shock on sperm cell membranes, in *Effects of Low Temperature on Biological Membranes* (Morris, G. J. and Clarke, A., eds.), Academic, London, pp. 189–218.
27. Hammerstedt, R. H., Graham, J. K., and Nolan, J. P. (1990) Cryopreservation of mammalian sperm: what we ask them to survive. *J. Andrology* **11,** 73–88.
28. Hofmo, P. O. and Almlid, T. (1990) Recent developments in freezing of boar semen with special emphasis on cryoprotectants, in *Boar Semen Preservation II: Proceedings of the Second International Conference on Boar Semen Preservation* (Johnson, L. A. and Rath, D., eds.), Paul Parey, Beltsville, MD, pp. 111–122.
29. Amann, R. P. and Pickett, B. W. (1987) Principles of cryopreservation and a review of cryopreservation of stallion spermatozoa. *Equine Vet. Sci.* **7,** 145–173.

CHAPTER 20

Cryopreservation of Mammalian Embryos

Slow Cooling

Jocelyn E. Hunter

1. Introduction

The benefits of banking mammalian embryos for long-term storage have long been recognized *(1)*. Cryopreservation of gametes prevents genetic drift, offering a viable alternative to maintaining active breeding colonies while safeguarding the genetic integrity of scientifically valuable strains. This is of particular importance in the fields of mouse genetics and in the conservation of endangered species. It has been suggested that embryo cryopreservation will be critical for the preservation of endangered species if the genetic diversity of the small populations in captive breeding programs is to be maintained *(2)*. The availability of stored genetic information in the form of embryos increases the generation time and allows the contribution of genetic material past the normal reproductive lifespan of the animal. The advances in the cryopreservation of domestic and laboratory animal embryos suggests that as knowledge of the basic reproductive physiology of exotic species increases, the establishment of embryo banks will become a reality *(2,3)*. In addition, the ease with which frozen embryos can be transported is of great benefit, overcoming the necessity to transport live animals while reducing the likelihood of disease transmission.

The survival of single-cell fertilized rabbit ova following the addition of 15% (v/v) glycerol and cooling to –15°C was first described by Smith in 1952 *(4)*. However, it was observed that subsequent cooling to –79°C significantly reduced the survival rate. In 1971 it was reported that eight-cell and blastocyst stage mouse embryos developed after rapid cooling in 7.5% (v/v) polyvinyl pyrrolidine to –79°C *(5)*. However, embryos could not be stored for longer than 30 min, and using the same method, two cell embryos failed to recover. It was not until the concurrent reports by Whittingham et al. *(6)* and Wilmut *(7)*, in which live births were achieved from transfer of frozen/thawed eight-cell mouse embryos, that the successful low temperature storage of mammalian embryos was realized. Both protocols were in agreement that optimal survival for the mouse embryo was attained using the permeating cryoprotectant dimethyl sulfoxide (DMSO) equilibrated at low suprazero temperatures and slow controlled rates of cooling and warming.

This basic protocol was used to successfully cryopreserve rat *(8)* and rabbit *(9)* embryos, whereas a modification in the rate of thawing resulted in the successful preservation of cow *(10)*, sheep *(11)*, and goat *(12)* embryos. Since then, cryopreservation of many mammalian species has become both a reliable and routine procedure and can yield pregnancy rates comparable to those achieved for nonfrozen embryo transfer. However, the main emphasis for cryopreservation techniques has remained centered around mouse and bovine embryos. Indeed, the mouse remains the most commonly used animal system for developing cryopreservation protocols. The interest in banking mouse strains is primarily a result of the use of mouse as a model for human disease and the use of mouse for techniques requiring early cleavage stage embryos, such as nuclear transfer and gene injection.

Many factors are known to influence survival during cryopreservation, including the cryoprotective additive (CPA), the concentration of the CPA, the rate of temperature decrease, the final temperature to which the cells are cooled, and the warming rate. Although the use of a CPA is necessary to obtain survival of embryos cooled to $-20°C$ or below, the method of addition and removal are flexible *(13,14)*. The termination temperature of the slow controlled rate cooling prior to plunging into liquid nitrogen for long-term storage dictates the optimum rate of warming. Fast thawing rates usually maximize survival when used in conjunction with a high subzero termination temperature ($> -40°C$), whereas slow cooling to $<-60°C$ is most successful when cells are thawed at slow controlled warming rates. It is well documented that in cryopreservation

Table 1
Cryopreservation of Mammalian Species

Species	Refs.
Antelope	23
Baboon	24,25
Bovine	10,14,26
Equine	17,27–29
Goat	12,30
Gorilla	31
Hamster	32
Monkey	33–36
Mouse	6,18,37,38
Ovine	11,39,40
Porcine	41,42
Rabbit	9,43,44
Rat	8,45

of preimplantation embryos, the various embryonic stages exhibit differences in their susceptibility to freezing damage, and the choice of the CPA is determined by the developmental stage *(15–17)*. The age and size of the embryo is also influential in determining the postthaw survival rate *(17,18)*. Research in cryopreservation of mouse embryos has shown that the success of a cryopreservation protocol will be significantly influenced by genotype *(19–21)*. It has been suggested that the paternal genotype *(20)* and both the maternal and embryonic genotype *(22)* can affect survival rates postthaw. It appears that all these factors interact to determine cryosurvival.

As previously stated, methods for freezing many preimplantation stages, from the pronucleate ova to the expanded blastocyst, have been published. However, particular stages, such as eight-cell for mouse and blastocyst for bovine preservation, are preferred and references for the different species and stages given in Table 1 will reflect this. The three methods outlined in this chapter are not comprehensive but provide a framework of the main protocols and cryoprotectants in use and have been chosen for the relative simplicity of cryoprotectant addition/removal. In addition, the methods in Section 3. are those used for mouse embryos; for details of protocols used for other species *see* Table 1.

2. Materials
2.1. Equipment

1. A controlled rate freezing machine can be obtained from several sources including Planer (Planer Products Ltd., Middlesex, UK) or Custom Biogenic Systems (Custom Biogenic Systems, New Baltimore, MI).
2. A heat sealer, such as Impulse Sealer (Consolidated Plastics, Twinsburg, OH).
3. 2-mL Plastic cryotubes (Nunc Cryotubes, Roskilde, Denmark).
4. Clear plastic straws (150-µL, Instruments de Medicine Veterinaire [IMV], l'Aigle, France).
5. 1-mL Monoject syringe.
6. Petri dishes (Falcon, Los Angeles, CA; Becton Dickinson, Oxnard, CA).
7. Water bath.
8. Liquid nitrogen dewars.
9. Safety equipment: cryogloves, goggles, cryoapron, and so on.
10. A 37°C incubator.

2.2. Chemicals

All chemicals used are Analar grade in purity.

1. Sucrose (Aldrich, Milwaukee, WI) used as a diluent.
2. Bovine serum albumin (BSA, purity 98% minimum, ICN Biochemicals, High Wycombe, UK). All cryoprotectants are made in phosphate-buffered saline (PBS) supplemented with 3–4 mg/mL BSA.
3. DMSO (Sigma-Aldrich, St. Louis, MO).
4. Glycerol (Aldrich).
5. Propanediol (propylene glycol, 1,2-propanediol, PROH, Sigma).

3. Methods
3.1. Cryopreservation Using DMSO as Cryoprotectant

The original method used to cryopreserve eight-cell mouse embryos using DMSO *(6,7)* is still used by many laboratories for preserving embryos. This is a method very similar to the original slow cool/slow thaw protocol with some minor alterations (*see* Notes 1–3).

1. Collect embryos and transfer into 0.1 mL of PBS supplemented with 3–4 mg/mL BSA in a 2-mL plastic tube. Store the cryotubes on ice at 0°C for a minimum of 30 min.
2. After 30 min add 0.1 mL of 2.0M DMSO to produce a final concentration of 1.0M DMSO.
3. After a 15 min equilibration period transfer the vials to an ice bath at –6°C and induce seeding (*see* Note 4).

4. Incubate the ampules at the seeding temperature in a controlled rate freezer for 5 min to allow dissipation of the latent heat of ice crystallization.
5. Cool the samples at a slow controlled rate (0.5°C/min) to –80°C. (Rates between 0.3 and 0.5°C/min all have been used successfully with this method.)
6. Plunge the samples directly into liquid nitrogen for long-term storage and note the location in the holder for ease of recovery at thawing.
7. Thawing can either be done using fast or slow controlled rate warming:
    a. Remove the cryotubes from the liquid nitrogen and hold vertical at room temperature, ensuring that no part of the vial is in contact with a surface. Surface contact will result in variable warming rates within the tube and alter the survival rates. Then dilute the samples from the cryoprotectant by slow addition (0.1 mL/min) of 0.8 mL of PBS to prevent the embryos from undergoing osmotic shock.
    b. Alternatively, transfer the tubes to a controlled rate freezer precooled to –110°C and warmed at 4°C/min to 0°C, and add a total of 0.8 mL of PBS over a 2-min period.
8. Locate the embryos and transfer to media for either embryo transfer or culture.

3.2. Cryopreservation Using Glycerol as Cryoprotectant

Glycerol has been used extensively as a cryoprotectant for cattle embryos, but has also been used for mouse embryos at various stages. Because of the fact that the embryo has a low permeability to glycerol and requires long equilibration periods, glycerol is usually added at room temperature or above. Permeability is known to decrease with decreasing temperature and thus equilibration times will be shorter at higher temperatures. This method is based on the protocol developed for two cell mouse embryos *(37)* *(see* Notes 1–3).

1. Collect the embryos in PBS, then incubate 20 embryos per vial equilibrated with 0.1 mL of 0.25M glycerol for 20 min in plastic cryotubes in a 37°C incubator.
2. Add a second 0.1-mL aliquot of cryoprotectant (1.0M glycerol) at 37°C and return the vials to the incubator for another 40 min equilibration period.
3. Transfer the vials to a programmable freezer set at 37°C and cool at 2°C/min to –7°C.
4. Induce ice nucleation and then hold the ampules at –7°C for 10–15 min to allow dissipation of the latent heat of ice crystallization *(see* Note 4).
5. Cool the samples at 0.3°C/min to –35°C at which point plunge directly into liquid nitrogen and transfer for storage at –196°C.

6. Thaw the samples rapidly by gentle agitation in a 37°C water bath until the ice has completely melted.
7. The cryoprotectant can be removed from the embryos either stepwise or in a single step:
   a. Remove the cryoprotectant stepwise. Using a two-step dilution, incubate the samples in CPA + $0.7M$ sucrose for 10–15 min followed by $0.7M$ sucrose for another additional 10–15 min (*see* Note 5).
   b. Alternatively, remove the embryos from the cryotubes and immediately transfer into a 60-mm Petri dish containing approx 3.0 mL of diluent. Expose the embryos to the diluent alone in a single step for 30 min, then transfer into PBS and then either into culture media for on-growth or transfer to pseudopregnant females.

3.3. Cryopreservation Using Propanediol as Cryoprotectant

Historically propanediol was rarely used for the preservation of embryos; DMSO and glycerol being the cryoprotectants of choice. Lassalle et al. *(46)* developed a protocol for the preservation of human embryos and this has been widely used and modified for use with other species and stages of embryo.

1. Preload clear plastic straws by fixing the straw to the tip of a 1-mL Monoject syringe, then aspirate the cryoprotectant into the straw using the syringe. Mark the straw approx 1 cm from the unplugged end. Place the tip of the straw into the diluent then aspirate to the 1-cm mark. Introduce a small air bubble and then aspirate a second column of diluent. Wipe the tip of the straw to prevent contamination of the CPA and draw the cryoprotectant up to the 1-cm marking. Introduce a second air bubble followed by a larger volume of cryoprotectant (approx 3 cm) into the straw. Draw a final air bubble into the straw followed by cryoprotectant to approx 0.5 cm from the tip (*see* Notes 6, 7, and 9–11).
2. Either leave the straw attached to the syringe to prevent leakage of the solution or detach and lay it horizontally, ready for loading the embryos.
3. Equilibrate the embryos for 30 min in $1.0M$ PROH at room temperature in a 35-mm Petri dish (*see* Note 8).
4. Load the straws with a maximum of 10 embryos/straw and seal immediately, take care to avoid dislodging the embryos from the column of cryoprotectant (*see* Note 8).
5. Place the straws in a controlled rate freezer and cool to –7°C at 2°C/min.
6. Induce ice nucleation by touching the straw with the tips of a pair of forceps cooled in liquid nitrogen (*see* Note 11).

7. After a holding period of 10 min, cool the straws at 0.3°C/min to −30°C and then rapidly cool to −190°C at 100°C/min or plunge directly into liquid nitrogen for storage.
8. To thaw, expose the straws to room temperature for 30–40 s, then incubate at 30°C in a water bath until the ice has completely melted.
9. Cut the ends of the straws and expel the contents into a 60-mm Petri dish, locate the embryos, and transfer through a series of 5-min washes of $1.0M$ PROH + $0.1M$ sucrose, $0.5M$ PROH + $0.1M$ sucrose, $0.1M$ sucrose, and finally into PBS then transfer for culture.

4. Notes

1. Embryos designated for low temperature storage may be of poor quality since they are frequently the excess embryos remaining after a transfer procedure. Even when all the embryos obtained from a collection are to be frozen, it is important to note that the quality of the embryos cryopreserved has a significant effect on the survival rate, the implantation rate, and thus the ability to attain live births *(47)*.
2. When attempting to establish an embryo bank, several factors must be considered to determine the number of embryos that are required to be frozen before the strain can be assumed to be successfully cryopreserved. As stated in Section 1., the genotype has a significant effect on survival rates postthaw. In addition, it is known that genotype affects the embryo quality and this in turn will affect survival of the cryopreservation procedure. In mouse embryo banking it is essential to consider the gene of interest, as some strains will be maintained in a heterozygous stock owing to the deleterious effects of the homozygous form. If the parents are both heterozygous, only 75% of the embryos will be carriers of the gene, however, only 50% would be carriers if the gene was a recessive lethal. In addition, the level of penetrance should be considered, since, depending on the background, this can vary and the numbers of embryos required to successfully recover the strain will be greater by temperature, oocyte age, and cumulus removal.

    It is advisable to thaw embryos and to recover liveborn before removing stocks from active breeding. A final consideration is that not all embryos from any one particular animal or strain should be stored in a single storage vessel to avoid the loss of all the genetic information in the event of a failure of the container to maintain temperature.
3. The developmental stage of the embryos is influential in achieving successful cryopreservation. Embryos at the exponential stage of development (two-, four-, and eight-cell stages) appear to be more resistant to damage during freezing and thawing than those at an intermediate cell division

stage (three-, five-, six-, and seven-cell stages). This is probably because the embryo is in nuclear arrest with the nuclear material confined within the nuclear membrane. Freezing is usually recommended during interphase to avoid chromosome abnormalities as a result of spindle damage owing to depolymerization of the temperature sensitive microtubule components *(38,48–50)*.
4. Although automated ice nucleation is now an option in many of the controlled rate freezers, the trend in the field still appears to be to seed the samples independently. Ice crystals can be induced using either a precooled tool, such as forceps cooled in liquid nitrogen, or by the introduction of a crystal from a pipet tip and observing ice formation in the individual samples.
5. There is a difference in density between diluent and CPA causing the embryos to rapidly disperse if transferred directly into diluent. Therefore, on thawing, it is advisable to remove the cryoprotectant by transferring the embryos into a combination of CPA and diluent, making handling and location of the embryos easier.
6. If straws are the chosen method for storing the embryos, a practical consideration is that owing to the thermal properties of straws, care must be taken to prevent the straws from warming up during handling. While allowing rapid rates of cooling and warming, straws are also susceptible to temperature changes during transfer from the controlled rate freezer to liquid nitrogen for long-term storage and during handling for purposes of identification prior to thawing.
7. Plastic straws have the advantage of allowing rapid dilution immediately on thawing while the embryos are still in the straw. It has also been suggested that freezing in straws may offer some protection against damage to the zona, such as zona fracture, although this may be a more important form of injury for faster rates of cooling than those described in Section 3.
8. The number of embryos equilibrated at one time is dictated by the skill of the handler. However, a maximum of 60 embryos should be equilibrated in cryoprotectant prior to transfer to straws, since handling time in the loading of the embryos into the straws can be significant, particularly if new to the technique.
9. When loading straws it is advisable that a small volume of CPA be aspirated in advance of the column of the CPA into which the embryos are loaded to prevent contamination. The sandwich arrangement used in many protocols of diluent, air, CPA, air, CPA + embryos, air, and CPA is advantageous. If the straw explodes on thawing the embryos are less likely to be lost.

10. The presence of media at either end of the straw separated from the embryos by air acts as a buffer against damage from excessive temperature increases during the heat sealing process.
11. When inducing seeding in straws it is advisable to touch the media at the opposite end to the embryos and to allow the ice crystals to propagate through the media. This prevents a rapid decrease in the temperature and a deviation from the optimal cooling rate.

References

1. Polge, C. (1977) The freezing of mammalian embryos: perspectives and possibilities, in *The Freezing of Mammalian Embryos* (Elliot, K. and Whelan, J., eds.), Elsevier, Amsterdam, pp. 3–13.
2. Ballou, J. D. (1992) Potential contribution of cryopreserved germ plasm to the preservation of genetic diversity and conservation of endangered species in captivity. *Cryobiology* **29,** 19–25.
3. Benirschke, K. (1984) The frozen zoo concept. *Zoo. Biol.* **3,** 311–323.
4. Smith, A. U. (1952) Behaviour of fertilized rabbit eggs exposed to glycerol and to low temperatures. *Nature* **170,** 374,375.
5. Whittingham, D. G. (1971) Survival of mouse embryos after freezing and thawing. *Nature* **233,** 125,126.
6. Whittingham, D. G., Leibo, S. P., and Mazur, P. (1972) Survival of mouse embryos frozen to –196°C and –296°C. *Science* **178,** 411–414.
7. Wilmut, I. (1972) The effect of cooling rate, cryoprotective agent and stage of development on survival of mouse embryos during freezing and thawing. *Life Sci.* **11,** 1071–1079.
8. Whittingham, D. G. (1975) Survival of rat embryos after freezing and thawing. *J. Reprod. Fertil.* **43,** 575–578.
9. Whittingham, D. G. and Adams, C. E. (1976) Low temperature preservation of rabbit embryos. *J. Reprod. Fertil.* **47,** 269–274.
10. Wilmut, I. and Rowson, L. E. A. (1973) The successful low temperature preservation of mouse and cow embryos. *J. Reprod. Fertil.* **33,** 352,353.
11. Willadsen, S. M., Polge, C., Rowson, L. E. A., and Moor, R. M. (1974) Preservation of sheep embryos in liquid nitrogen. *Cryobiology* **11,** 560.
12. Bilton, R. J. and Moore, N. W. (1976) In vitro culture, storage and transfer of goat embryos. *Aust. J. Biol. Sci.* **29,** 125–129.
13. Critser, J. K., Arneson, B. W., Aaker, D. V., Huse-Benda, A. R., and Ball, G. D. (1988) Factors affecting the cryosurvival of mouse two-cell embryos. *J. Reprod. Fertil.* **82,** 27–33.
14. Niemann, H. (1991) Cryopreservation of ova and embryos from livestock: current status and research needs. *Theriogenology* **35,** 109–124.
15. Wilmut, I., Polge, C., and Rowson, L. E. A. (1975) The effect on cow embryos of cooling to 20, 0 and –196°C. *J. Reprod. Fertil.* **45,** 409–411.
16. Polge, C., Wilmut, I. A., and Rowson, L. E. A. (1974) The low temperature preservation of cow, sheep, and pig embryos. *Cryobiology* **11,** 560.

17. Wilson, J. M., Kramer, D. C., Potter, G. D., and Welsh, T. H. (1986) Nonsurgical transfer of cryopreserved equine embryos to pony mares treated with exogenous progestin. *Theriogenology* **25,** 227.
18. Van der Auwera, I., Cornillie, F., Pijnenborg R., and Konickx, P. R. (1992) The age of pronucleate mouse ova influences their development in vitro and survival after freezing. *Hum. Reprod.* **7,** 660–665.
19. Schneider, U. and Maurer, R. R. (1983) Factors affecting survival of frozen-thawed mouse embryos. *Biol. Reprod.* **29,** 121–128.
20. Schmidt, P. M., Hansen, C. T., and Wildt, D. E. (1985) Viability of frozen-thawed mouse embryos is affected by genotype. *Biol. Reprod.* **32,** 507–514.
21. Schmidt, P. M., Schiewe, M. C., and Wildt, D. E. (1987) The genotypic response of mouse embryos to multiple freezing variables. *Biol. Reprod.* **37,** 1121–1128.
22. Pomp, D. and Eisen, E. J. (1990) Genetic control of survival of frozen mouse embryos. *Biol. Reprod.* **42,** 775–786.
23. Kramer, L., Dresser, B. L., Pope, R. D., Dalhausen, R. D., and Baker, R. D. (1983) The nonsurgical transfer of frozen-thawed Eland *(Tragelaphus oryx)* embryos. *Annu. Proc. Am. Assoc. Zoo. Vet.* 104–195.
24. Pope, C. E., Pope, V. Z., and Beck, L. R. (1984) Live birth following cryopreservation and transfer of a baboon embryo. *Fertil. Steril.* **42,** 143–145.
25. Pope, C. E., Pope, V. Z., and Beck, L. R. (1986) Cryopreservation and transfer of baboon embryos. *J. In Vitro Fert. Embryo Transfer* **3,** 33–39.
26. Fahning, M. L. and Garcia, M. A. (1992) Status of cryopreservation of embryos from domestic animals. *Cryobiology* **29,** 1–18.
27. Slade, N. P., Takeda, T., Squires, E. L., and Elsden, R. P. (1985) A new procedure for the cryopreservation of equine embryos. *Theriogenology* **24,** 45–57.
28. Reiger, D., Bruyas, J. F., Lagneaux, D., Bezard, J., and Palmer, E. (1991) The effects of cryopreservation on the metabolic activity of day 6.5 horse embryos. *J. Reprod. Fertil. Suppl.* **44,** 411–417.
29. Wilson, J. M., Caceci, T., Potter, G. D., and Kraemer, D. C. (1987) Ultrastructure of cryopreserved horse embryos. *J. Reprod. Fertil.* **35(Suppl.),** 405–417.
30. Rong, R., Guangya, W., Jufen, Q., and Jianchen, W. (1989) Simplified quick freezing of goat embryos. *Theriogenology* **31,** 252.
31. Lanzendorf, S. E., Holmgren, W. J., Schaffer, N., Hatasaka, H., Wentz, A. C., and Jeyendran, R. S. (1993) In vitro fertilization and gamete micromanipulation in the lowland gorilla. *J. Assist. Reprod. Genet.* **9,** 358–364.
32. Iida, T. (1992) The effect of cryopreservation on early development and chromosome constitution in Chinese hamster embryos. *Asia Oceania J. Obstet. Gynaecol.* **18,** 407–412.
33. Balmaceda, J. P., Heitman, T. O., Garcia, M. R., Pauerstein, C. J., and Pool, T. B. (1986) Embryo cryopreservation in cynomolgus monkeys. *Fertil. Steril.* **45,** 403–406.
34. Summers, P. M., Shephard, A. M., Taylor, C. T., and Hearn, J. P. (1987) The effects of cryopreservation and transfer on embryonic development in the common marmoset monkey, *Callithrix jacchus. J. Reprod. Fertil.* **79,** 241–250.
35. Wolf, D. P., Vandevoort, C. A., Meyer-Haas, G. R., Zelinski-Wooten, M. B., Hess, D. L., Baughman, W. L., and Stouffer, R. L. (1989) *In vitro* fertilization and embryo transfer in the rhesus monkey. *Biol. Reprod.* **41,** 335–346.

36. Lanzendorf, S. E., Zelinski-Wooten, M. B., Stouffer, R. L., and Wolf, D. P. (1990) Maturity at collection and the developmental potential of rhesus monkey oocytes. *Biol. Reprod.* **42,** 703–711.
37. Bernard, A. and Fuller, B. J. (1983) Cryopreservation of in vitro fertilized 2 cell mouse embryos using a low glycerol concentration and normothermic cryoprotectant equilibration: a comparison with in vivo fertilized ova. *Cryo-Lett.* **4,** 171–177.
38. Balakier, H., Zenzes, M., Wang, P., MacLusky, N. J., and Casper, R. F. (1991) The effect of cryopreservation on the development of S- and G2-phase mouse embryos. *J. In Vitro Fertil. Embryo Transfer* **8,** 89–95.
39. Heyman, Y., Vincent, C., Garnier, V., and Cognie, Y. (1987) Transfer of frozen-thawed embryos in sheep. *Vet. Rec.* **120,** 83–85.
40. McGinnis, L. K., Duplantis, S. C., Jr., Waller, S. L., and Youngs, C. R. (1989) The use of ethyl glycol for cryopreservation of sheep embryos. *Theriogenology* **31,** 226.
41. Kashiwazaki, N., Ohtani, S., Miyamoto, K., and Ogawa, S. (1991) Production of normal piglets from hatched blastocysts frozen at –196 degrees C. *Vet. Rec.* **128(11),** 256,257.
42. Fujini, Y., Ujisato, Y., Endo, K., Tomizuka, T., Kojima, T., and Oguri, N. (1993) Cryoprotective effect of egg yolk in cryopreservation of porcine embryos. *Cryobiology* **30,** 299–305.
43. Al-Hasani, S., Hepnar, C., Diedrich, K., van der Ven, H., and Krebs, D. (1992) Cryopreservation of rabbit zygotes. *Hum. Reprod.* **7(Suppl. 1),** 81–83.
44. Vicente, J. S. and Garcia-Ximenez, F. (1993) Effects of strain and embryo transfer model (embryo from one versus two donor does/recipient) on results of cryopreservation in rabbit. *Reprod. Nutr. Dev.* **33,** 5–13.
45. Stein, A., Fisch, B., Tadir, Y., Ovadia, J., and Kraicer, P. F. (1993) Cryopreservation of rat blastocysts: a comparative study of different cryoprotectants and freezing/thawing methods. *Cryobiology* **30,** 128–134.
46. Lassalle, B., Testart, J., and Renard, J. P. (1985) Human embryo features that influence the success of cryopreservation with the use of 1,2 propanediol. *Fertil. Steril.* **44,** 645–651.
47. Kennedy, L. G., Boland, M. P., and Gordon, I. (1983) The effect of embryo quality at freezing on subsequent development of thawed cow embryos. *Theriogenology* **19,** 823–832.
48. Chedid, S., Van den Abbeel, E., and Van Steirteghem, A. C. (1992) Effects of cryopreservation on survival and development of interphase- and mitotic-stage 1-cell mouse embryos. *Hum. Reprod.* **7,** 1451–1456.
49. Ng, S. C., Sathananthan, A. H., Wong, P. C., Ratnam, S. S., Ho, J., Mok, H., and Lee, M. N. (1988) The fine structure of early human embryos frozen with 1,2-propanediol. *Gamete Res.* **19,** 253–263.
50. Testart, J., Lassalle, B., Forman, R., Gazengel, A., Belaisch-Allart, J., Hazout, A., Rainhorn, J. D., and Frydman, R. (1987) Factors influencing the success rate of human embryo freezing in an in vitro fertilization and embryo transfer program. *Fertil. Steril.* **48,** 107–112.

CHAPTER 21

Cryopreservation of Mammalian Embryos

Vitrification

Magosaburo Kasai

1. Introduction

In 1972, Whittingham et al. (*1*) reported the first successful deep-freezing of mouse embryos. The method was efficient and reproducible, and it has been widely used since. This method includes a slow cooling process (this lasts a few hours after ice seeding). Recent attempts to improve the method have focused on shortening the cooling process. Vitrification, as reported by Rall and Fahy in 1985 (*2*), reduces the cooling stage duration to a minimum. Samples can be plunged directly into liquid nitrogen from temperatures above 0°C. An additional advantage of this approach is that high levels of viability of the embryos may be maintained; this is primarily attributed to the absence of extracellular ice during freezing.

Vitrification is defined as "the solidification of a solution brought about not by crystallization but by an extreme elevation in viscosity during cooling" (*3*). To induce vitrification in liquid nitrogen, a solution must contain very high concentrations of cryoprotective agent(s). For successful cryopreservation of mammalian embryos by vitrification, these cryoprotectants must permeate the cell membrane and be concentrated intracellularly to avoid lethal intracellular ice formation. However, during exposure to the solution, the embryos are liable to be injured by the toxicity of the high concentration of the cryoprotectant(s), this being the greatest obstacle to successful vitrification. Strategies to avoid this toxic effect include lowering the concentration of the cryoprotectant and lowering the temperature. Rall and Fahy *(2)* equilibrated embryos in a solution containing low concentrations of cryoprotectants at room temperature, and then transferred the embryos to a concentrated vitrification solution at a low temperature (4°C). However, this stepwise treatment requires a long time (>30 min), and treatment of embryos at 2 different temperatures is less practical. In addition, the initial vitrification solution (named VS1) was composed of relatively toxic ingredients including, dimethyl sulfoxide (DMSO), acetamide, and propylene glycol. Other strategies include: the use of less toxic cryoprotectant(s), and reducing the exposure time. Recently, the author's group composed a new vitrification solution with lower toxicity, which enables one-step exposure of embryos at a room temperature before rapid cooling *(4)*. In this chapter, the author describes the procedure of a vitrification method developed for mouse morulae, which results in virtually 100% postthaw survival.

2. Materials

2.1. Solutions

1. PB1 medium: Dulbecco's phosphate-buffered saline (PBS), modified by addition of 5.56 mM glucose, 0.33 mM pyruvate, 100 IU/mL penicillin, and 3 mg/mL bovine serum albumin (BSA) *(5)*. Store at 4°C after filtration through a 0.2-μm filter.
2. Sucrose solution: PB1 medium containing 0.5M sucrose. Store at 4°C after filtration through a 0.2-μm filter (*see* Note 1).
3. EFS (ethylene glycol, ficoll, and sucrose) solution: PB1 medium containing 40% (v/v) ethylene glycol, 18% (w/v) Ficoll, and 0.3M sucrose *(4)* (*see* Notes 1 and 2). Filtration is possible with a 0.45-μm filter. Store at 4° or –20°C. To make this solution, first prepare FS solution, which is PB1 medium containing 30% (w/v) Ficoll, and 0.5M sucrose, then mix it with ethylene glycol.

    The precise method is as follows: To prepare the FS solution, add 3.51 mL of filter sterilized PB1 medium (omitting the BSA) and 1.5 g Ficoll 70 (Pharmacia, Uppsala, Sweden), mol wt 70,000 to a 10-mL plastic test tube with a tight stopper. Mix and dissolve Ficoll thoroughly. Add 0.856 g sucrose and dissolve it thoroughly. Then add 10.5 mg BSA and dissolve.

    To prepare the EFS solution: Add ethylene glycol and FS solution together in the ratio of 2:3, to make 40% (v/v) ethylene glycol, 18% (w/v) Ficoll, and 0.3M sucrose, in a 10-mL test tube with a tight stopper. Use 1-mL disposable syringes (with an 18-gage needle), to measure solution vol-

umes, as the viscosity of the liquids will make accurate pipeting impossible. Mix completely by repeated tilting.
4. M 16: A modified Krebs-Ringer bicarbonate solution *(6)*.
5. Liquid paraffin.
6. Liquid nitrogen.

2.2. Equipment

1. Tightly stoppered 10-mL plastic test tube.
2. 0.25-mL Insemination straws (A-201, Istruments de Medicine Veterinaire [IMV], l'Aigle, France).
3. Straw connector (2-cm long silicone tube that fits the straw).
4. Straw powder.
5. Dewar flask.
6. Pasteur pipet.
7. 1-mL Disposable syringe, 18-gage needle.
8. Embryological watch glass.
9. Timer.
10. A 37°C incubator.
11. A water bath.
12. Safety equipment: cryogloves, goggles, face mask, and so on.

3. Method

A standard method for mouse morulae is described (for other species and stages, *see* Notes 3–5).

1. Control of the temperature: Adjust the room temperature. Temperature on the bench should be 20°C ($\pm$ 0.5°C). Equilibrate all the solutions and instruments at 20°C (*see* Note 6).
2. Loading of EFS solution into the straw: Connect a 1-mL syringe and a 0.25-mL straw with a silicone tube. By aspirating the syringe carefully, load 55-mm sucrose solution, 25 mm air, 4-mm EFS solution, 4 mm air, and 13-mm EFS solution, successively, into the straw as shown in Fig. 1, and place it horizontally near the edge of the bench (*see* Note 7).
3. Exposure of embryos to EFS solution: Pour approx 0.2 mL of EFS solution into a watch glass. Under a dissecting microscope, pick up about 10 embryos in PB1 medium at the tip of a pipet, suspend the embryos in the EFS solution with a minimal volume of PB1 medium, and start the timer. The embryos shrink instantly and move up to the surface of the EFS solution. Dispel the PB1 medium from the pipet and aspirate the EFS solution and then embryos into the pipet. Transfer the embryos to other part of the EFS solution in the watch glass to wash embryos in EFS solution (removal

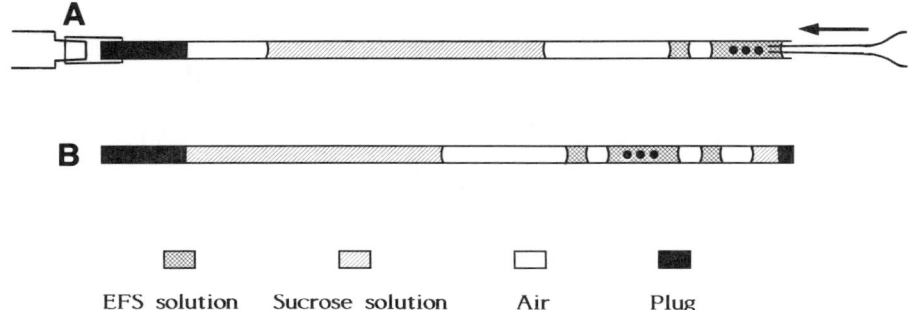

Fig. 1. Configuration of the straw. (**A**) Introduction of embryos (•) into the EFS solution and (**B**) the straw after sealing.

of PB1 medium). When embryos are not being manipulated under the microscope, keep the lights off so as not to elevate the temperature of the microscope stage. The total exposure time of the mice morulae should not exceed 2 min at 20°C (*see* Note 6).
4. Loading of embryos into the straw and sealing: Pick up the embryos in the pipet and transfer them into the 13-mm EFS column in the straw. By successive aspiration of the syringe, load 4 mm air, 4 mm EFS, and then air until the first larger column of sucrose solution comes close to the cotton plug. Finally, aspirate sucrose solution completely to seal the cotton plug end. Pull the straw out of the connector and put the open end into the straw powder, then place in a beaker containing water at 20°C to seal. Avoid warming the EFS columns with fingers (*see* Notes 7–9).
5. Cooling and preservation: Pour liquid nitrogen into a dewar flask to a depth of 5–6 cm. (This should be prepared beforehand.) Two minutes after exposure of the embryos to EFS solution, plunge the straw so that half of the straw containing the embryos is in liquid nitrogen, the rest of the straw being in liquid nitrogen vapor. Leave it for at least 1 min until the sucrose solution freezes. Then store the straw in the liquid nitrogen tank (*see* Note 10).
6. Warming and recovery of embryos: Place 0.3 mL of sucrose solution and PB1 medium under paraffin oil in a culture dish. Fill a 1-mL syringe with an 18-gage needle attached, with 1 mL of sucrose solution. Remove the straw from the liquid nitrogen storage tank using forceps, immerse it into a 20°C water bath, and shake gently. In about 10 s, when the sucrose solution begins to melt, remove the straw and wipe the water quickly with a paper towel. Keeping the straw horizontal, hold it at the larger sucrose

column end, cut off the powder plug and then the cottonwool plug. Tilt the straw (30°) with EFS side down, and perfuse the straw slowly with 1-mL sucrose solution in the syringe into a watch glass (*see* Note 11). After perfusion, leave sucrose solution in the straw and keep it horizontal. Start the timer and gently shake the watch glass to mix the solution. Under a dissecting microscope, recover the embryos in a pipet with the sucrose solution (the sucrose causes the embryos to shrink within the zona), then transfer them to the sucrose solution in a culture dish. If all the embryos are not recovered, decant the residual sucrose solution remaining in the straw into the watch glass and examine under a dissecting microscope. Transfer the embryos into PB1 medium 5 min after perfusion (*see* Note 12).

7. Assessment of embryonic viability: Wash the embryos by transferring them to fresh PB1 medium. Check for any physical injury in the zona pellucida and cytoplasm. Culture the embryos in 0.3 mL of M 16 in a CO_2 incubator at 37°C. Physical injury of blastomeres is clearly observable after 1–2 h of culture. After 48 h incubation, more than 95% of the embryos will develop to expanded blastocysts. These embryos will develop to term if transferred to recipient animals.

4. Notes

1. The roles of Ficoll and sucrose: Macromolecules facilitate vitrification of solutions *(3)*. As a macromolecule, polyethylene glycol was used in the first report *(2)*. However, Ficoll seems to have advantages of low toxicity, high solubility, and low viscosity. As an alterative, Schiewe et al. *(7)* used BSA, but it is more expensive.

    Sucrose does not penetrate the cell, and it exerts a considerable osmotic effect, inducing the cells to shrink. Dehydration of the cell facilitates intracellular vitrification, and also prevents osmotic swelling during removal of permeated cryoprotectants from the cells. By adding 0.3*M* sucrose to PB1 medium containing 40% (v/v) ethylene glycol + 18% (w/v) Ficoll, the toxicity of the solution is reduced significantly *(4)*, presumably because sucrose reduces the total amount of intracellular ethylene glycol by reducing the cell volume.

2. Ethylene glycol: To avoid lethal intracellular ice formation, permeation of an agent into the cell is essential. However, since the concentration of the cryoprotectant(s) is extremely high in vitrification solutions, full permeation is not necessary as it causes injury owing to chemical toxicity of the cryoprotectants, or osmotic stress during removal. Of the three additives in the EFS solution, only ethylene glycol can permeate the cell. Ethylene glycol was selected as a low toxic agent among major permeating

cryoprotective agents, such as glycerol, DMSO, propylene glycol, and acetamide. However, the high concentration of ethylene glycol is responsible for the toxicity of EFS solution.
3. Embryos at stages other than the morula: The standard protocol described here was developed for mouse morulae (ICR strain) *(4)*. The compact morula has the advantage of affording the highest survival after vitrification among embryos from one-cell to expanded blastocyst stage *(8)*. The standard procedure is also suitable for eight-cell mouse embryos and early blastocysts, without significant reduction in survival. One-cell, two-cell, and four-cell embryos can also be vitrified by the standard method, although in vitro developmental rates are slightly lower than those of untreated controls (71, 87, and 88%, respectively) *(8)*. Because the permeability of the embryo to a cryoprotectant changes considerably with its developmental stage, the optimum time and temperature for exposure of the embryo of other stages before vitrification may not be the same as just described. For example, the survival of blastocysts vitrified by the standard method decreases as the blastocoel enlarges. Blastocysts with a large blastocoel should be treated in two steps at 25°C before cooling. First suspend the embryos in 10% (v/v) ethylene glycol in PB1 medium at 25°C for 5 min, then load into the EFS column in the straw with a minimal volume of the solution, and plunge into liquid nitrogen after 30 s of exposure to EFS solution. Using this procedure, more than 90% of expanded blastocysts reexpand during culture *(9)*.
4. Vitrification of oocytes: This protocol is unsuitable for unfertilized mouse oocytes. Nakagata *(10)* reported successful vitrification of mouse oocytes using a procedure in which oocytes are exposed to a vitrification solution similar to VS1 for only 5–10 s prior to rapid cooling.
5. Embryos of species other than the mouse: In the rabbit, the standard method for mouse morulae is effective for morulae and early blastocysts. For in vivo development, rabbit embryos must have an intact zona pellucida; as little damage occurs using the standard protocol, high survival rates may be attained *(11)*. In cattle, blastocysts produced through maturation, fertilization and development in vitro have been vitrified after one-step exposure to EFS solution; 72% of the recovered embryos could reexpand in vitro, and had the ability to develop to term after transfer to recipient cows *(12)*.
6. Exposure time and temperature: The permeability of embryos to a cryoprotectant is closely related to the temperature. It also affects both the degree of cryoprotection and toxic injury. Therefore, optimum time for exposing embryos to EFS solution is largely dependent on the ambient temperature; the higher the temperature, the shorter the exposure time *(13)*.

This standard protocol employs a 2-min exposure of mouse morulae to EFS solution at 20°C. However, at this temperature a 30-s exposure also results in high survival rates after vitrification. In this case, the embryos in PB1 medium in a pipet should be directly introduced into the EFS column in the straw, without prewashing in EFS solution in a watch glass. Embryos must be loaded at the tip of a fine pipet, with a slightly larger diameter than the embryo to minimize the volume of PB1 medium introduced into EFS solution. Because the straw must be sealed within 30 s, the last small EFS column can be omitted. It has also been reported that mice morulae can survive up to 5 min of exposure to EFS solution at 20°C before rapid cooling (4). However, a 2-min exposure time is recommended as excess permeation is not required.

At 25°C, exposure time of mouse morulae to EFS solution should be only 30 s rather than 2 min by direct introduction of embryos into the straw, as just described. When mouse morulae are manipulated in a cold room, high survival is maintained after exposure for 2–5 min at 10°C, or for 2–10 min at 5°C, before cooling. Because the permeability of the embryo to a cryoprotectant varies with the developmental stage, optimum conditions for treatment (exposure time and temperature) of embryos at other stages than the morula may not be the same as described (*see* Notes 3 and 5).

7. Configuration of the straw: The first column of sucrose solution in the straw helps dilute EFS solution after warming. The sucrose column also indicates melting time during warming. The first small EFS column removes the sucrose solution sticking to the inner surface of the straw. Contamination of the second EFS column, which is for embryo loading, with sucrose solution can reduce survival. The third EFS column separates embryos and sucrose solution, and may trap embryos in case they should stick to the inner wall during moving of the column. However, if it is difficult to seal the straw within the required time, this column may be omitted. The EFS solution is confined to a rather small portion of the straw in order to increase the dilution rate with the sucrose solution at perfusion after warming. The last sucrose column is necessary for powder sealing. This column is small in order to minimize the distance the EFS column moves (*see* Fig. 1).

8. Container: Alternatively, it is possible to vitrify embryos in cryotubes. Pipet embryos into a small amount (<100 µL) of EFS solution in the cryotube. Immediately after warming add sucrose solution (1 mL) to the tube and mix quickly to dilute EFS solution.

9. Straw sealing: Using this protocol straw sealing was achieved using powder (IMV) because it is quick to use and does not heat up the straw. However, heat sealing using a sealer is also possible, assuming the embryos in the

EFS column are not heated. If the sealing is incomplete, liquid nitrogen will penetrate into the straw and cause the straw to rupture and, consequently, loss of the embryos.

10. Cooling: If the whole straw is immersed in liquid nitrogen in one step, the straw will rupture because of the rapid increase in the volume of the freezing sucrose solution. An alternative method of cooling in a large volume of liquid nitrogen is to hold the straw using forceps and immerse EFS side of the straw rapidly (*see* Fig. 1), followed by slow immersion of the rest (10 s).

    A few percent of vitrified embryos preserved using the standard protocol have been observed to have a damaged zona pellucida. The damage may have resulted from fracture of the glassy suspension caused by the very rapid cooling rate (2400°C/min, between 0 and −150°C). Ice crystals may be formed if a vitrification solution is cooled too slowly. However, Schiewe et al. *(7)* cooled embryos by suspending them in liquid nitrogen vapor. It is also possible to cool EFS solution in a straw in cold nitrogen vapor without crystallization.

11. Warming: The vitrified solution must be warmed rapidly to prevent crystallization (devitrification), which can be injurious. By direct transfer of the straw from liquid nitrogen into 20°C water, warming occurs at 1600°C/min (between −150 and 0°C), and ice is not formed in the EFS solution. After warming, the straw must be perfused quickly, because prolonged exposure of embryos to warmed EFS solution will lead to reduced viability.

12. Removal of cryoprotectants: Ethylene glycol permeated into the cell is removed in the sucrose solution, which promotes the shrinkage of the embryo. However, immediately after suspension in sucrose solution, embryos may swell osmotically, because the osmolality in the cell is higher than that of the sucrose solution, and because water penetrates the cell more rapidly than ethylene glycol diffuses out. Although the mouse morulae vitrified by this protocol are not injured, this hypotonic stress, in general, can cause injury not only in vitrification but also in conventional freezing protocols.

References

1. Whittingham, D. G., Leibo, S. P., and Mazur, P. (1972) Survival of mouse embryos frozen to −196°C and −269°C. *Science* **178,** 411–414.
2. Rall, W. F. and Fahy, G. M. (1985) Ice-free cryopreservation of mouse embryos at −196°C by vitrification. *Nature* **313,** 573–575.
3. Fahy, G. M., MacFarlane, D. R., Angell, C. A., and Meryman, H. T. (1984) Vitrification as an approach to cryopreservation. *Cryobiology* **21,** 407–426.
4. Kasai, M., Komi, J. H., Takakamo, A., Tsudera, H., Sakurai, T., and Machida, T. (1990) A simple method for mouse embryo cryopreservation in a low toxicity

vitrification solution, without appreciable loss of viability. *J. Reprod. Fertil.* **89,** 91–97.
5. Whittingham, D. G. (1971) Survival of mouse embryos after freezing and thawing. *Nature* **233,** 125,126.
6. Hogan, B., Costantini, F., and Lacy, E. (1986) *In vitro* culture of eggs, embryos and teratocarcinoma cells, in *Manipulating the Mouse Embryos: A Laboratory Manual* (Hogan, B., Costantini, F., and Lacy, E., eds.), Cold Spring Harbor Laboratory, Cold Spring Harbor, NY, pp. 245–267.
7. Schiewe, M. C., Rall, W. F., Stuart, L. D., and Wildt, D. E. (1991) Analysis of cryoprotectant, cooling rate and in situ dilution using conventional freezing or vitrification for cryopreserving sheep embryos.*Theriogenology* **36,** 279–293.
8. Miyake, T., Kasai, M., Zhu, S. E., Sakurai, T., and Machida, T. (1993) Vitrification of mouse oocytes and embryos at various stages in an ethylene glycol-based solution by a simple method. *Theriogenology* **40,** 121–134.
9. Zhu, S. E., Kasai, M., Otoge, H., Sakurai, T., and Machida, T. (1993) Cryopreservation of expanded mouse blastocysts by vitrification in ethylene glycol-based solutions. *J. Reprod. Fertil.* **98,** 139–145.
10. Nakagata, N. (1989) High survival rate of unfertilized mouse oocytes after vitrification. *J. Reprod. Fertil.* **87,** 479–483.
11. Kasai, M., Hamaguchi, Y., Zhu, S. E., Miyake, T., Sakurai, T., and Machida, T. (1992) High survival of rabbit morulae after vitrification in an ethylene glycol-based solution by a simple method. *Biol. Reprod.* **46,** 1042–1046.
12. Tachikawa, S., Otoi, T., Kondo, S., Machida, T., and Kasai, M. (1993) Successful vitrification of bovine blastocysts, derived by in vitro maturation and fertilization. *Mol. Reprod. Dev.* **34,** 266–271.
13. Kasai, M., Nishimori, M., Zhu, S. E., Sakurai, T., and Machida, T. (1992) Survival of mouse morulae vitrified in an ethylene glycol-based solution after exposure to the solution at various temperatures. *Biol. Reprod.* **47,** 1134–1139.

Chapter 22

Cryopreservation of Human Gametes

Jocelyn E. Hunter

1. Introduction

Spermatozoa were the first mammalian cells to be successfully cryopreserved *(1)*, and the techniques developed by Sherman *(2)* enabled centers to begin to establish sperm banks for men at risk of losing their fertility as a result of clinical treatments for illness. Banks of cryopreserved donor sperm have become an integral part of assisted reproduction procedures, allowing the treatment of couples with male factor infertility by ensuring that samples are available when required while conferring anonymity. However, more recently it has become essential that human cells for donation are quarantined to reduce the risks of contamination, and semen is required to be stored for several months to ensure the requisite tests for HIV are performed. Although pregnancies from both in vitro fertilization (IVF) and intrauterine insemination (IUI) are common, cryopreservation causes cell lysis, a reduction in motility, an increase in the number of cells undergoing spontaneous acrosome reaction, an alteration in the distribution of acrosin *(3,4)*, and an overall decrease in fertility *(5)*. Sublethal cryodamage is apparent in postthaw analysis of donor sperm. Spermatozoa may be initially intact and motile, however, viability postthaw is lost rapidly over time. This may be a result of cold shock damage, although it appears that human sperm are less susceptible than many mammalian species, possibly as a result of the high sterol content of the membrane, since high cholesterol levels stabilize membranes during cooling *(6)*. In addition, there is an increase in lipid peroxidase damage leading to an alteration in the metabolic state *(7)* and the sperm are exposed to hyperosmotic stress during freezing and thawing that is related to cryoinjury *(8)*. Attempts have been made to compare freezing procedures and to optimize the protocols used *(9,10)*, however there appears to be no concurrence on the protocol yielding the highest survival rates.

Although mammalian embryos were first cultured outside the maternal uterus in 1890 *(11)*, it was not until 1939 that the unfertilized oocyte could be successfully matured in vitro *(12)*. With the introduction of laparoscopy, these techniques were applied to humans, preovulatory oocytes were collected, matured to the point of polar body extrusion *(13)*, fertilized, and cultured in vitro to the hatching blastocyst stage *(14)*. This was followed by the development to term and live birth of an oocyte fertilized after collection following natural cycle folliculogenesis *(15)*. Exogenous endocrine stimulation was introduced, achieving multiple follicular development while preventing spontaneous ovulation. However, the recovery of numerous preovulatory oocytes and subsequent successful fertilization resulted in excess embryos. Although replacing more than a single embryo enhances the likelihood of implantation, it is accompanied by an increased risk of multiple pregnancies *(16)*. Cryopreservation of these excess embryos for replacement in subsequent unstimulated cycles can overcome the problems associated with the production of multiple embryos *(17)*. Since the first successful report of a live birth following the transfer of a frozen/thawed eight-cell embryo *(18)*, the long-term storage of human embryos has raised both ethical and legal problems *(19,20)*.

Low temperature storage of the unfertilized oocyte is an attractive alternative to embryo preservation overcoming many of the ethical and legal objections associated with human embryo storage. In addition, oocyte cryopreservation would be beneficial for those females about to undergo chemo/radiotherapy allowing in vitro fertilization on recovery. Although births following the transfer of frozen thawed embryos are now common, pregnancies following the cryopreservation of the mature unfertilized human oocyte are infrequent. Initial reports of successful fertilization, implantation, and development to term following freezing and thawing of human oocytes were promising *(21–24)*. The technique is not sufficiently reliable to offer as a clinical option.

Indeed exposure to low suprazero temperatures has been associated with a reduction in fertilization *(25,26)*. In addition, exposure to cryoprotectants alone can reduce the fertilization rate *(27,28)*. Reduced rates of fertilization may be a result of the premature release of cortical granules causing zona hardening *(29)* or alternatively as a result of depolymerization of microtubules in the meiotic spindle and chromosome dispersal from the equatorial plate *(30)*. This may result in an increased risk of aneuploidy *(16,31,32)*. Subcortical microfilaments have been shown to be sensitive to dimethyl sulfoxide (DMSO) and propanediol however, this appears to be sensitive to both the temperature and time of exposure. Although it has been reported that the mature unfertilized oocyte has been successfully cryopreserved using propanediol and sucrose *(33)*, propanediol has also been shown to increase polyspermy *(21)* and cause parthenogenetic activation *(34)*.

It appears that in a clinical situation the preferred stage for embryo cryopreservation is the pronucleate ova using propanediol as the cryoprotective additive. In addition, although the rate of warming is variable, the cooling rates chosen most often are slow cooling rate of approx 0.3–0.5°C/min. Few reports of successful vitrification or rapid freezing *(35–37)* have been published and the techniques are not widely used for clinical purposes, being associated with potential toxicity or chromosome abnormalities *(37,38)*. For the unfertilized ova the literature tends to conclude that the success rates are too low to offer the procedure as a viable clinical treatment. The methods described in this chapter are not comprehensive but provide the most commonly used protocols for slow or rapid freezing of human embryos and sperm.

2. Materials
2.1. Materials for Embryo Cryopreservation
1. Gametes and embryos to be frozen will be those not fertilized or transferred in the intial replacement during the IVF procedure.
2. Sterile plastic Petri dishes (Falcon, Los Angeles, CA; Becton Dickinson, Oxnard, CA).
3. Sterile pipets/micropipets.
4. 37°C CO_2 incubator.
5. Culture media: The culture media used routinely in the unit should be used for the final wash medium when embryos are thawed, embryos should be wash twice in culture media prior to transfer to normal IVF culture conditions for assessment of survival and assessment of suitability for transfer. Many centers use Earle's Balanced Salt Solution (EBSS, Gibco, Gaithersburg, MD) supplemented with approx 10–20% sera, either human serum albumin or heat-denatured patients serum.

6. All cryoprotectant solutions are prepared using phosphate-buffered saline (PBS, Gibco) supplemented with serum, either human serum albumin or heat-denatured patients serum.
7. Diluent: PBS or sucrose (Aldrich, Milwaukee, WI) in PBS.
8. Cryoprotectant (CPA): DMSO (Sigma-Aldrich, St. Louis, MO), glycerol (Aldrich), and propanediol (propylene glycol, 1,2-propanediol, PROH, Sigma). All chemicals are Analar grade purity.
9. A minimum of two long-term storage vessels, a dewar suitable for use with liquid nitrogen for precooling forceps for seeding and transporting samples to and from the cryogenic storage containers.
10. Forceps (*see* Notes 4 and 11 in Chapter 20 on ice nucleation).
11. Protective equipment: eye goggles, gloves, and apron.
12. A controlled-rate freezing machine, such as those offered by Planar (Middlesex, UK) or Custom Biogenic Systems (New Baltimore, MI).
13. Depending on the method employed, there is a choice of freezing receptacle of either:
    a. Cryovials, such as 2-mL plastic cryotubes (Nunc Cryotubes, Roskilde, Denmark) or
    b. Clear plastic straws (150 µL) (Instruments de Medicine Veterinaire [IMV], l'Aigle, France).
14. If straws are used, they can be sealed using commercially available plugs (obtained from company supplying straws), a heat sealer, such as Impulse Sealer (Consolidated Plastics, Twinsburg, OH), or by plugging the ends with powder (IMV) that solidifies when wet.
15. 1-mL Monoject syringe for preloading straws with cryoprotectant and diluent (*see* Notes 9–11 in Chapter 20 for details concerning loading of straws).

2.2. Materials for Sperm Freezing

1. Egg yolk buffer can be purchased commercially as Tris and TES test egg yolk buffer (Irvine Scientific, Irvine, CA) or prepared freshly: Add 15% (v/v) glycerol (Aldrich), 20% (v/v) egg yolk, 1.15 g sodium citrate, 1.8 g glycine, 1000 IU penicillin, and 50 mg of streptomycin to 1 L of distilled water. All chemicals are Analar grade purity. Several additional methods for the preparation of egg yolk buffer can be found in the literature *(39–41)*.
2. A minimum of two long-term storage vessels.
3. A controlled-rate freezing machine (as detailed in Section 2.1., item 12).
4. Safety equipment (as detailed in Section 2.1., item 11).
5. Clear plastic insemination straws (1.5 mL, IMV) that can be sealed using plugs obtainable from IMV, a heat sealer, or sealant powder.

3. Methods
3.1. Methods for Embryo Cryopreservation (see Note 1)
3.1.1. Pronucleate Zygotes

This method has been successful using both straws and cryovials and, although most commonly used for pronucleate embryos, this procedure has been used for early cleavage stage embryos *(42)*. Using vials, pregnancy rates identical to those achieved using fresh embryos have been reported *(43)* (*see* Notes 2 and 3).

3.1.1.1. CRYOVIALS

1. Load the cryovials with 0.3 mL of 1.5M PROH in PBS, transfer the embryos into the tubes, and leave to equilibrate at room temperature for 30 min.
2. Transfer the tubes to a controlled-rate freezing machine and cool at 1°C/min to between –6 and –7°C; maintain at this temperature for 5 min. Then manually seed the vials by touching the outside of the vial with precooled forceps at the level of the meniscus, until ice crystals can be visualized.
3. Hold the samples at –6 to –7°C for another 5 min to allow dissipation of the latent heat of crystallization before cooling at 0.5°C/min to –80°C, then plunge directly into liquid nitrogen for storage.
4. Thaw the vials by transferring the tubes to a programmable freezing machine precooled to –100°C and warm at 8°C/min to room temperature.
5. Then transfer the embryos through a serial dilution of 1.0M PROH, 0.5M PROH, and finally PBS; incubate for 5 min in each solution.
6. Assess the embryos immediately for morphological damage/survival and then culture under CO_2 at 37°C for 24 h prior to assessment for transfer (*see* Notes 4–8).

3.1.1.2. STRAWS

This method is also suitable for use with straws *(44)*.

1. Equilibrate the embryos for 15 min in 1.0M PROH in PBS supplemented with 20% human cord serum (HCS) at room temperature in a 35-mm Petri dish.
2. Preload clear plastic straws using a 1-mL Monoject syringe to aspirate the cryoprotectant, 1.5M PROH + 0.2M sucrose (in PBS + 20% HCS).
3. Load the straws with a single embryo, place in a controlled-rate freezer, and cool to –7°C at 2°C/min (*see* Note 9).
4. Induce ice nucleation by touching the straw with the tips of a pair of forceps cooled in liquid nitrogen.
5. Cool the straws at 0.3°C/min to –30°C then cool rapidly (100°C/min) to –190°C. Transfer directly into liquid nitrogen for storage.

6. Thaw the embryos by exposing the straws to room temperature for 30–40 s, followed by agitation in a 30°C water bath until the ice has melted.
7. Locate the embryos and transfer through a series of 5-min washes in 1.5M PROH + 0.2M sucrose, 1.0M PROH + 0.2M sucrose, 0.5M PROH + 0.2M sucrose, 0.2M sucrose, PBS, and finally transfer to culture medium for 24 h (*see* Notes 4–8).

3.1.2. Early Cleavage Stage Embryos

Although PROH has been used with early cleavage stage embryos, DMSO is usually the cryoprotectant chosen *(18,45)* (*see* Note 10).

1. Prepare solutions of DMSO, 0.75M and 1.5M, in either PBS or HEPES-buffered T6 media supplemented with 10% (v/v) heat-denatured patients serum immediately prior to the cryopreservation procedure.
2. Equilibrate the embryos at room temperature with 0.75M DMSO for 10 min, then transfer the embryos into individual ampules containing 0.3 mL of 1.5M DMSO at 0°C and equilibrate for another 10 min (*see* Note 11).
3. Cool the vials in a programmable freezer at 2°C/min to –6°C and then induce ice crystallization with precooled forceps. Hold the samples at –6°C for 10 min, cool at 0.3°C/min to –30°C, then plunge directly into liquid nitrogen.
4. Thaw samples slowly by warming at 8°C/min, then wash and transfer for culture (*see* Notes 4 and 5).

3.1.3. Blastocyst Stage Embryos

For later stage embryos, glycerol and slow cooling rates are preferred. Owing to the low permeability of the embryo to glycerol, even at 37°C or room temperature, the method requires the stepwise addition and removal of the CPA *(46,47)*.

1. Transfer the embryos through increasing concentrations of 1, 2, 4, 6, and 8% (v/v) glycerol incubating for 10 min at each concentration to a final concentration of 10% (v/v) glycerol. This avoids direct osmotic shock to the embryos.
2. Transfer a single embryo into a freezing vial or straw containing 10% (v/v) glycerol at room temperature and place in a controlled-rate freezer equilibrated at room temperature.
3. Cool the vials at 1°C/min to –7°C and induce ice nucleation by touching the vial at the level of the meniscus using precooled forceps.
4. Maintain the vials at –7°C for 10 min to dissipate the heat of fusion, cool at 0.3°C/min to –36°C, then plunge into liquid nitrogen.

5. Thaw the vials rapidly by incubation in a 30°C water bath and agitate gently until the ice is completely melted.
6. Remove the cryoprotectant by stepwise washing in decreasing concentrations of glycerol, 8% (v/v) for 10 min, 6% (v/v) for 12 min, 5% (v/v) for 14 min, 4% (v/v) for 16 min, 3% (v/v) for 18 min, 2% (v/v) for 20 min, then wash twice in PBS.
7. Finally wash the embryos in culture media and assess for morphological survival (*see* Notes 4 and 5).

3.1.4. Rapid Freezing Techniques

The use of slow cooling rates in conventional cryopreservation requires the use of controlled-rate freezing machines which can be costly. In addition, the embryo is exposed to potentially damaging concentrations of CPAs for long periods of time at high subzero temperatures. It would be advantageous to have available rapid, simple techniques, such as those used for the vitrification or ultrarapid freezing of mouse embryos *(48,49)*, for the cryopreservation of human embryos. Although mouse embryos have been successfully cryopreserved using these techniques, they are not widely applied to human material *(36)*. The following method has been reported to achieve successful live births *(35)*.

1. Prepare a solution of $4.5M$ DMSO + $0.3M$ sucrose in PBS supplemented with 20% (v/v) serum up to 24 h prior to use and store at 4°C (*see* Note 9).
2. Pipet single embryos directly into clear plastic straws loaded with precooled cryoprotectant solution, seal the straws, and plunge directly into liquid nitrogen.
3. Thaw the embryos rapidly by immersion in a 37°C water bath and wash in PBS + $0.3M$ sucrose for 6 min. Then wash the embryos in culture media and examine for survival (*see* Notes 4 and 5).

3.2. Methods for Sperm Freezing

Glycerol is frequently employed as a cryoprotectant in the cryopreservation of human spermatozoa at concentrations in the range of 10–20% (v/v) *(1,2,4,40,50)*. Research groups and medical units differ in whether they use straws or ampules, glycerol independently or combined with an egg yolk buffer, commercially purchased test egg yolk buffer or freshly made egg yolk buffer, or whether a controlled-rate freezer or liquid nitrogen vapor is used to cool the sperm. The following methods are those cited most frequently in the literature. For all the methods used, the initial collection and analysis procedure is standard.

1. Collect donor sperm according to the unit's own regulations. It is usually recommended that the samples should be provided after a 3-d abstinence period and the analysis conform to the World Health Organizations guidelines.
2. Allow the sample to liquify for up to 1 h at 37°C, analyze, and mix well with the cryoprotectant solution (*see* Note 12).

3.2.1. Glycerol

1. Add glycerol to the sample to give a final concentration of 7.5% (v/v). Transfer 0.5-mL aliquots into cryovials or draw into straws.
2. Transfer the vials/straws to a controlled-rate freezing machine and cool from room temperature to 0°C at 1°C/min.
3. Cool the samples from 0°C to –100°C at 10°C/min, plunge into liquid nitrogen, and store.
4. Thaw samples at room temperature for 30 min and then perform complete semen analysis.
5. If the sample is to be used for donor insemination (DI), no further preparation is usually required, however, if the sample is to used for either an IUI or an IVF, attempt to remove the cryoprotectant. Centrifuge the sample at 200g for approx 6–10 min, resuspend the pellet in culture medium, then centrifuge for a second 10-min period. Overlay the resulting pellet with culture media, allow the sperm to swim up for 30 min, then adjust the final concentration to approx 1×10^6/mL for insemination (*see* Notes 13–16).

3.2.2. Glycerol–Egg Yolk Buffer

1. Dilute the semen in a 1:1 ratio with Tris and TES test egg yolk buffer (Irvine Scientific) or an egg yolk buffer freshly prepared in the laboratory (*see* Section 2.2.).
2. Draw 0.5 mL of the semen/cryoprotectant mixture into each straw and transfer them to a controlled-rate freezer at room temperature, then cool at 1.7°C/min to –6°C.
3. Then cool the samples at 5°C/min to –100°C and then transfer to liquid nitrogen for storage.
4. Thaw samples at room temperature for 30 min until the ice has completely melted then analyze semen and preparation for use for insemination (*see* Section 3.2.1., step 5).

3.2.3. Vapor Freezing

Semen can be cryopreserved successfully without the necessity of using controlled cooling rates.

Cryopreservation of Human Gametes 229

1. As in Section 3.2.2., dilute the semen in a 1:1 ratio with Tris and TES test egg yolk buffer (Irvine Scientific) or an egg yolk buffer freshly prepared in the laboratory (*see* Section 2.2.).
2. Draw 0.5 mL of the semen/cryoprotectant mixture into each straw, then transfer to a refrigerator and incubate at 4°C for 1 h (*see* Note 12).
3. Place the samples in the vapor phase for 5 min, then plunge into liquid nitrogen for storage.
4. Thaw samples at room temperature for 30 min until the ice has completely melted, then analyze semen and preparation for use for insemination (*see* Section 3.2.1., step 5).

4. Notes

1. Human embryos should be frozen in individual vials or straws to avoid the possibility of the loss of larger numbers if the straw or vial were to be damaged or exposed to suboptimal freezing conditions. In addition, this allows greater flexibility during thawing, since if any of the initial embryos thawed fail to survive a replacement can be thawed easily.
2. The storage receptacle used depends on both the method of cryopreservation and the stage of embryo that is to be routinely frozen, and this must be taken into account when purchasing long-term storage vessels. Briefly, straws offer the benefit of being sealed easily and thus unlikely to be contaminated; however, the heat transfer properties of straws means they are much quicker to warm up on handling. This allows faster rates of cooling and warming to be achieved, however, it may be problematic when straws are required to be handled for identification purposes. Although cryotubes are less sensitive to temperature fluctuations, they are more susceptible to leakage into the vial and the possibility of contamination. Tubes can be sealed to reduce the likelihood of this occurring (Cryoflex Sealant, Nunc).
3. Liquid phase storage ensures the samples are subject to fewer temperature variations and that the system can be maintained at a constant temperature in the event of a power failure. Alarmed vessels have the added precaution that levels of liquid nitrogen can be monitored and provide a warning in the event of the failure of a vessel. It is recommended that if automated filling techniques are used, the levels should be checked manually at regular intervals. In addition, a minimum of two independent vessels should be used to protect from the possibility of the loss of all of a patient's samples in the event of a failure of the vacuum system.
4. The quality of the embryos to be frozen is of extreme importance in determining the survival rate on thawing. This is discussed in Chapter 20, Notes 1 and 3.

5. Embryo quality should be assessed both before and immediately after cryopreservation, since this is influential in determining the survival rate and whether an embryo is suitable for replacement. Pronucleate and early cleavage stage embryos are suitable for cryopreservation if they have normal refractive cytoplasm, an intact zona pellucida, few cytoplasmic fragments, and regular-sized blastomeres. In addition, they should either have only two pronuclei visible at approx 16–22 h postinsemination or have undergone the correct number of cell divisions for the time in culture. Expanded blastocysts should have reached the expanded stage by no later than d 6 of culture, have an intact zona, and a clearly visible blastocoele and inner cell mass. On thawing, embryos should have an intact zona, refractive cytoplasm with no vacuoles, and a minimum of 50% of the blastomeres intact. Early cleavage stage and pronucleate embryos should undergo cleavage and expanded blastocysts should reexpand, reforming the blastocoele in overnight culture.
6. Although cryopreservation at the pronucleate stage results in high survival rates, there is no point at which the embryo quality can be judged prior to freezing, and this may result in too few or only poor quality embryos for transfer. The possible reduction in the number of embryos available at the 48-h designated transfer time may decrease the possibility of establishing a successful pregnancy. In extreme cases when none of the remaining embryos divide, the patient may have no embryos to transfer.
7. Embryo survival on thawing can only be judged by the ability of the pronucleate ova to enter into cell division. To ensure that an embryo transfer occurs, more embryos than can be replaced may need to be thawed or fewer than optimal replaced.
8. Recent reports using human and mouse pronucleate stage embryos have shown that there is a higher survival rate if embryos are frozen in the G2 phase, or non-DNA synthesis phase (*see* Note 3 in Chapter 20). Indeed, it has been suggested that pronucleate human ova may be best frozen at 20–22 h postinsemination but only when the two pronuclei are still clearly visible. This becomes more complicated for multicellular stage embryos that have asynchronous cycles during cell division and thus the optimal time is not as obvious.
9. The presence of sucrose in the CPA solution in which the embryo is to be frozen during the equilibration phase of the procedure has the benefit of partially dehydrating the embryo and reducing the chance of intracellular ice formation as the temperature is reduced.
10. Improved culture techniques, such as the use of feeder cell layers, have resulted in an increase in the quality of embryos, increasing the number of cells in the embryo if cultured to the blastocyst *(51)*. This may be benefi-

cial for patients who require genetic analysis of embryos since cells may be biopsied, the embryo cultured until analysis is completed, and, following transfer, the remaining unaffected embryos frozen at later stages. Since the cryopreserved embryos may be of a higher quality, the survival rates may increase.

11. Equilibrating embryos and oocytes with DMSO at room temperature may damage the tubulin-containing microtubules or actin-containing microfilaments of the cytoskeleton. It has been reported that DMSO can cause depolymerization of both microfilaments and microtubules *(27,28)*, thus equilibration with DMSO is usually recommended at approx 4°C or below.

12. It has been suggested that holding semen for longer than 1 h prior to freezing is detrimental to sperm survival *(41)*.

13. The substantial emotional and monetary investment required for an IVF treatment makes a failed cycle owing to poor quality donor sperm unsatisfactory. All donors should be required to undergo an initial assessment of sperm survival of the cryopreservation procedure used. Tests routinely used to determine sperm quality, such as semen analysis, hamster egg penetration test, cervical mucus penetration, and the acrosome status should be performed on the sample postthaw and the donor assessed for suitability for cryopreservation.

14. Centrifugation can result in a decrease in the number of motile sperm and damage to sperm head membrane. This may cause the release of products, such as free radicals from the dead or damaged sperm, which can be potentially damaging to the remaining viable sperm.

15. The speed at which the sperm are centrifuged, and therefore the force to which they are exposed, can be reduced by centrifuging in the presence of a Percol gradient of 0.3 mL of increasing concentrations of 50, 70, and 95% (v/v) Percol. The majority of the damaged sperm are removed at the concentration boundaries, whereas the motile sperm are concentrated in a pellet that can be easily resuspended in culture media and analyzed to give an optimum concentration of motile sperm for insemination.

16. The addition of pentoxifylline, 2-deoxyadenosine, or caffeine stimulates motility through alterations in the influx of Ca^+, which stimulates membrane-bound adenyl cyclase and increases the intracellular cAMP, causing an increase in motility and the number of acrosome reacted sperm. However, there is concern over the possible mitogenic side effects of the use of these chemicals to improve the quality of cryopreserved sperm. An alternative to these procedures may be to use platelet activating factor (PAF), which is found to occur naturally in semen and to have similar effects on addition *(50)*.

References

1. Polge, C., Smith, A. U., and Parkes, A. S. (1949) Revival of spermatozoa after vitrification and dehydration at low temperatures. *Nature* **164,** 666.
2. Sherman, J. K. (1954) Freezing and freeze drying of human spermatozoa. *Fertil. Steril.* **5,** 357–371.
3. Cross, N. L. and Hanks, S. E. (1991) Effects of cryopreservation on human sperm acrosomes. *Hum. Reprod.* **6,** 1279–1283.
4. Mahadevan, M. and Trounson, A. O. (1984) Effect of cooling, freezing and thawing rates and storage conditions on preservation of human spermatozoa. *Andrologia* **16,** 52–60.
5. Byrd, W., Bradshaw, K., Carr, B., Edman, C., Odom, J., and Ackerman, G. A. (1990) Prospective randomized study of pregnancy rates following intrauterine and intracervical insemination using frozen donor sperm. *Fertil. Steril.* **53,** 521–527.
6. Drobnis, E. Z., Crowe, L. M., Berger, T., Anchordoguy, T. J., Overstreet, J. W., and Crowe, J. H. (1993) Cold shock damage is due to lipid phase transitions in cell membranes—a demonstration using sperm as a model. *J. Exp. Zool.* **265,** 432–437.
7. Alvarez, J. G. and Storey, B. T. (1992) Evidence for increased lipid peroxidative damage and loss of superoxide dismutase activity as a mode of sublethal cryodamage to human sperm during cryopreservation. *J. Androl.* **13,** 232–241.
8. Gao, D. Y., Ashworth, E., Watson, P. F., Kleinhans, F. W., Mazur, P., and Critser, J. K. (1993) Hyperosomotic tolerance of human spermatozoa: separate effects of glycerol, sodium chloride, and sucrose on spermolysis. *Biol. Reprod.* **49,** 112–123.
9. Centola, G. M., Raubertas, R. F., and Mattox, J. H. (1992) Cryopreservation of human semen. Comparison of cryopreservatives, sources of variability, and prediction of post-thaw survival. *J. Androl.* **13,** 283–288.
10. Kolon, T. F., Philips, K. A., and Buch, J. P. (1992) Custom cryopreservation of human semen. *Fertil. Steril.* **58,** 1020–1023.
11. Heape, W. (1890) Preliminary note on the transplantation and growth of mammalian ova within a uterine foster mother. *Proc. Royal Soc. B* **48,** 457,458.
12. Pincus, G. and Saunders, B. (1939) The comparative behaviour of mammalian eggs in vivo and in vitro. VI The maturation of human ovarian ova. *Anat. Rec.* **75,** 537–545.
13. Edwards, R. G. (1965) Maturation in vitro of human ovarian oocytes. *Lancet* **vi,** 926–929.
14. Steptoe, P. C., Edwards, R. G., and Purdy, J. M. (1971) Human blastocysts grown in culture. *Nature* **229,** 132,133.
15. Steptoe, P. C. and Edwards R. G. (1978) Birth after reimplantation of a human embryo. *Lancet* **ii,** 366.
16. Edwards, R. G., Fishel, S. B., Cohen, J., Fehilly, C. B., Purdy, J. M., Slater, J. M., Steptoe, P. C., and Webster, J. M. (1984) Factors influencing the success of in vitro fertilization for alleviating human infertility. *J. In Vitro Fert. Embryo Transfer* **1,** 3–23.
17. Edwards, R.,G. and Steptoe, P. C. (1977) The relevance of the frozen storage of human embryos. *Ciba Found. Symp.* **52,** 235–250.

18. Trounson, A. O. and Mohr, L. (1983) Human pregnancy following cryopreservation, thawing and transfer of an eight-cell embryo. *Nature* **305,** 707–709.
19. Capron, A. M. (1992) Parenthood and frozen embryos. More than property and privacy. *Hastings Cent. Rep.* **22,** 32,33.
20. Perry, C. and Schneider, L. K. (1992) Cryopreserved embryos: who shall decide their fate? *J. Leg. Med.* **13,** 463–500.
21. Al-Hasani, S., Diedrich, K., van der Ven, H., Reinecke, A., Hartje, M., and Krebs, D. (1987) Cryopreservation of human oocytes. *Hum. Reprod.* **2,** 695–700.
22. Chen, C. (1986) Pregnancy after human oocyte cryopreservation. *Lancet* **i,** 884–886.
23. Chen, C. (1988) Pregnancy after human oocyte cryopreservation. *Ann. NY Acad. Sci.* **541,** 541–549.
24. Van Uem, J. F. H. M., Siebzehnrueble, E. R., Schuh, B., Koch, R., Trotnow, S., and Lang, N. (1987) Birth after cryopreservation of unfertilized oocytes. *Lancet* **i,** 752,753.
25. Bernard, A., Hunter, J. E., Fuller, B. J., Imoedemhe, D., Curtis, P., and Jackson, A. (1992) Fertilization and embryonic development of human oocytes after cooling. *Hum. Reprod.* **7,** 1447–1450.
26. Sathananthan, A. H., Trounson, A., Freeman, L., and Brady, T. (1988) The effects of cooling human oocytes. *Hum. Reprod.* **3,** 968–977.
27. Pickering, S. J., Braude, P. R., and Johnson, M. H. (1991) Cryoprotection of human oocytes: inappropriate exposure to DMSO reduces fertilization rates. *Hum. Reprod.* **6,** 142,143.
28. Trounson, A. and Kirby, C. (1989) Problems in the cryopreservation of unfertilized eggs by slow cooling in dimethyl sulfoxide. *Fertil. Steril.* **52,** 778–786.
29. Johnson, M. H., Pickering, S. J., and George, M. A. (1988) The influence of cooling on the properties of the zona pellucida of the mouse oocyte. *Hum. Reprod.* **3,** 383–387.
30. Johnson, M. H. and Pickering, S. J. (1987) The effect of dimethylsulphoxide on the microtubular system of the mouse oocyte. *Development* **100,** 313–324.
31. Glenister, P. H., Wood, M. J., Kirby, C., and Whittingham, D. G. (1987) Incidence of chromosome anomalies in first-cleavage mouse embryos obtained from frozen-thawed oocytes fertilized in vitro. *Gamete Res.* **16,** 205–216.
32. Kola, I., Kirby, C., Shaw, J., Davey, A., and Trounson, A. (1988) Vitrification of mouse oocytes results in aneuploid zygotes and malformed fetuses. *Teratology* **38,** 467–474.
33. Imoedemhe, D. G. and Sigue, A. B. (1992) Survival of human oocytes cryopreserved with or without the cumulus in 1,2-propanediol. *J. Assist. Reprod. Genet.* **9,** 323–327.
34. Shaw, J. M. and Trounson, A. O. (1989) Parthenogenetic activation of unfertilized mouse oocytes by exposure to 1,2-propanediol is influenced by temperature, oocyte age, and cumulus removal. *Gamete Res.* **24,** 269–279.
35. Feichtinger, W., Hochfellner, C., and Ferstl, U. (1991) Clinical experience with ultra-rapid freezing of embryos. *Hum. Reprod.* **6,** 735,736.
36. Trounson, A. and Sjoblom, P. (1988) Cleavage and development of human embryos in vitro after ultrarapid freezing and thawing. *Fertil. Steril.* **50,** 373–376.

37. Trounson, A., Peura, A., Freemann, L., and Kirby, C. (1988) Ultrarapid freezing of early cleavage stage human embryos and eight-cell mouse embryos. *Fertil. Steril.* **49,** 822–826.
38. Fahy, G. M., Lilley, T. H., Linsdell, H., Douglas, M. S., and Meryman, H. T. (1990) Cryoprotectant toxicity and cryoprotectant toxicity reduction: in search of molecular mechanisms. *Cryobiology* **27,** 247–268.
39. Mahadevan, M. and Trounson, A. O. (1983) Effect of cryoprotective media and dilution methods on the preservation of human spermatozoa. *Andrologia* **15,** 355–366.
40. Prins, G. S. and Weidel, L. (1986) A comparative study of buffer systems as cryoprotectants for human spermatozoa. *Fertil. Steril.* **46,** 147–149.
41. Yavetz, H., Yogev, L., Homonnai, Z., and Paz, G. (1991) Prerequisites for successful human sperm cryobanking: sperm quality and prefreezing holding time. *Fertil. Steril.* **55,** 812–816.
42. Troup, S. A., Matson, P. L., Critchlow, J. D., Morroll, D. R., Lieberman, B. A., and Burslem, R. W. (1990) Cryopreservation of human embryos at the pronucleate, early cleavage and blastocyst stages. *Eur. J. Obstet. Gynaecol. Reprod. Biol.* **38,** 133–139.
43. Veeck, L. L., Amundson, C. H., Brothman, L. J., DeScisciolo, C., Maloney, M. K., Muasher, S. J., and Jones, H. W., Jr. (1993) Significantly enhanced pregnancy rates per cycle through cryopreservation and thaw of pronuclear stage oocytes. *Fertil. Steril.* **59,** 1202–1207.
44. Testart, J., Lassalle, B., Forman, R., Gazengel, A., Belaisch-Allart, J., Hazout, A., Rainhorn, J. D., and Frydman, R. (1987) Factors influencing the success rate of human embryo freezing in an in vitro fertilization and embryo transfer program. *Fertil. Steril.* **48,** 107–112.
45. Camus, M., Van den Abbeel, E., Van Waesberghe, L., Wisanto, A., Devroey, P., and Van Steirteghem, A. C. (1989) Human embryo viability after freezing with dimethylsulfoxide as a cryoprotectant. *Fertil. Steril.* **51,** 460–465.
46. Cohen, J., Simons, R. F., Edwards R. G., Fehilly, C. B., and Fishel, S. B. (1985) Pregnancies following the frozen storage of expanding human blastocysts. *J. In Vitro Fertil. Embryo Transfer* **2,** 59–64.
47. Hartshorne, G. M., Elder, K., Crow, J., Dyson, H., and Edwards, R. G. (1991) The influence of in-vitro development upon post-thaw survival and implantation of cryopreserved human blastocysts. *Hum. Reprod.* **6,** 136–141.
48. Rall, W. F. and Fahy, G. M. (1985) Ice-free cryopreservation of mouse embryos at −196 degrees C by vitrification. *Nature* **313,** 573–575.
49. Rall, W. F., Wood, M. J., Kirby, C., and Whittingham, D. G. (1987) Development of mouse embryos cryopreserved by vitrification. *J. Reprod. Fertil.* **80,** 499–504.
50. Wang, R., Sikka, S. C., Veeraragavan, K., Bell, M., and Hellstrom, W. J. (1993) Platelet activating factor and pentoxifylline as human sperm cryoprotectants. *Fertil. Steril.* **60,** 711–715.
51. Menezo, Y., Nicollet, B., Herbaut, N., and Andre, D. (1992) Freezing cocultured human blastocysts. *Fertil. Steril.* **58,** 977–980.

CHAPTER 23

Cryopreservation of Human Red Blood Cells

Michael J. G. Thomas and Susan H. Bell

1. Introduction
1.1. The Red Cell

The red blood cell (RBC), or erythrocyte, is a flexible biconcave disc 8 µm in diameter. Its main function is to carry oxygen from the lungs to the tissues of the body and to perform this it contains a pigment, hemoglobin. During its 120-d lifespan, it travels about 300 miles around the arteriovenous circulation, repeatedly passing through the capillary bed. As the mean diameter of a capillary is about 3 µm, the red cell has to retain a high degree of flexibility which requires energy. Energy is generated as adenosine triphosphate (ATP) by the anaerobic, glycolytic pathway.

The shelf-life of blood, when stored in an optimal additive solution at 4°C, is 5 wk. There are circumstances in which it would be advantageous to store blood for a longer period. These include extension of shelf-life, the provision of transfusion material to individuals with rare blood groups, stockpiling against disaster, insurance against irregular supply, and the advoidance of infectious disease. Any attempt to prolong the shelf-life must provide a product that has all the aforementioned functions intact. One of the ways of extending the shelf-life is to freeze the red cells.

1.2. Background

There has been interest in the cryopreservation of erythrocytes since the 1940s, and a large range of compounds have been examined as possible cryoprotectants (1). Extracellular additives have included dextrans; polyvinylpyrrolidones (PVP); albumin; hydroxyethyl starch (HES); polyethylene oxide; polyethylene glycol; polyoxypropylene; polyglycol alcohols, such as mannitol and sorbitol; "Hemaccel;" detergents; and nonpenetrating sugars, such as lactose, maltose, sucrose, and dextrose. Intracellular additives included glycerol, dimethyl sulfoxide (DMSO), ethylene, diethylene, propylene glycols, acetamide, formamide, ethanol, methanol, monoacetin, and calcium lactobionate. In addition to this range of possible cryoprotectants, a number of other experimental variables have been examined. These include the quantity of red cells, the relative amount of cryopreservative added, and the rate of freezing and thawing.

From such experiments most cryoprotectants were abandoned as inefficient, toxic, or because only small quantities of cells could be cryopreserved; glycerol has been the cryoprotectant of choice.

In the 1960s, much work was done with extracellular (i.e., nonpenetrating) cryoprotectants, particularly dextrans of different molecular weights, and PVP. In 1967 Knorpp (cited in ref. 1) investigated the plasma expander HES as an extracellular cryoprotective agent; since then HES of differing molecular weights and degrees of substitution have been used in the cryopreservation of erythrocytes (2). One major problem with this method had been the requirement for very rapid freeze and thaw rates to avoid cell damage. Although small volumes of erythrocytes could be frozen rapidly, thereby producing relatively high yields, units containing the red cells from a full donation froze and thawed at a slower rate with a much greater amount of cellular destruction, resulting in a product that was of unacceptable quality for transfusion. The method patented by the Army Blood Supply Depot (ABSD) in 1990 has overcome this difficulty. Units containing all the red cells from a normal donation, frozen by this method, on thaw, give a recovery of 99% intact cells. After resuspension in plasma of the same blood group for .5 h, over 95% of the erythrocytes remain stable. The method, which is rapid, requires no sophisticated equipment, nor postthaw washing stage, is described in Section 3.2.

The five main methods of cryopreserving blood cells are listed in Table 1. The methods regularly used by the ABSD are described in fuller detail in the following sections.

Table 1
Cryopreservation of Human RBCs

No.	Method
1	High temperature/high glycerol method: Meryman and Hornblower *(3)*
2	High temperature/high glycerol method: Valeri *(4)*
3	Low temperature/low glycerol method—ABSD: Section 3.1.
4	HES method for whole donations—ABSD: Thomas *(5)*: Section 3.2.
5	HES method for 1–5-mL samples of whole blood—ABSD: Section 3.3.

2. Materials

2.1. RBC Cryopreservation Using Low Concentration Glycerol—ABSD

1. Gambro bags DF 1200 (Gambro, Sidcup, Kent, UK).
2. Air filter—Acrodisc (Gelman, Ann Arbor, MI) 0.2-µm filter.
3. Air Inlets, cat no. Travenol C0413 (Baxter Healthcare, Nr Newbury, Berks, UK).
4. 500-mL Bottles of cryoprotectant solution: 140 mL glycerol BP, 15 g Mannitol BP, 3.25 g NaCl BP, and water to 500 mL (available from Queen Elizabeth Medical Center, Birmingham, UK).
5. Leakproof plastic protective bags.
6. Frozen blood register.
7. Blood packs for cryopreservation.
8. Bottlehangar.
9. Plasma transfer set, Fenwal (Baxter) AU.8 Code No. 4C2243.
10. Fisons (Berlin, Germany) HAAKE SWB 20 rocking water bath.
11. Immersion bath containing an ethanol/distilled water/hibitane mixture in the ratio 75:15:10.
12. COBE/IBM (Quedgeley, Gloucs, UK) 2991 cell washing machine.
13. COBE/IBM processing set, consisting of doughnut bag with S tube attachment (912-647-819).
14. 40-µm Microaggregate blood filter (PALL SQ 403 [Portsmouth, UK]).
15. 1000 mL Red cell wash solution A (3.5% hypertonic saline solution) (Parkfields RSSU [Wolverhampton, UK]).
16. 1000 mL 0.9% NaCl iv infusion BP (Fresenius Polyfusor, Runcorn, Cheshire, UK).
17. Air inlet set (Avon A81 [Baxter]).
18. Beckman (High Wycombe, Bucks, UK) J6 centrifuge.
19. LNR40 liquid nitrogen refrigerator.

20. BPLD (Borehamwood, Hertfordshire, UK) freezing frame.
21. Safety equipment: cryogloves, goggles, and so on.

2.2. HES Method for Whole Donations—ABSD (see Note 1)

1. An LNR40 as the freezing vessel.
2. A thermostatically controlled circulating water bath set at 43.5°C.
3. An LNR450, or a mechanical freezer capable of achieving temperatures below −90°C, for the storage of the frozen product.
4. Polyimide/FEP freezing bags (NPBI, Amstelveen, The Netherlands).
5. A cryoprotectant pouch containing 40 mL of a 40% (w/v) solution of HES (Laevosan, Linz, Austria).
6. A white blood cell filter, with integral filtrate and satellite bag (PALL or NPBI).
7. Aluminium freezing frame and metal clips (ABSD or NPBI).
8. A refrigerated centrifuge. Beckman J6B, J6M, or J6ME.
9. A tube sealer modified to seal the "Nuplastic" tubing on the freezing bag (Sebra, Tucson, AZ).
10. Haemonetics Sterile Connecting Device SCD 312.
11. SCD cartridges (Haemonetics [Leeds, UK] 00325-00).
12. Hand tongs.
13. Colorimeter (CIBA Corning [Halstead, Essex, UK] Colorimeter Model 253).
14. Centrifuge (Centra 3S IEC Damon [Dunstable, Beds, UK]).
15. Diluter (Hamilton [Bonaduzag, Switzerland] Microlab-P).
16. Drabkins reagent.
17. Microhematocrit centrifuge.
18. Microhematocrit slide reader.
19. Citrate phosphate dextrose (CPD) or citrate phosphate dextrose with adenine (CPDA) blood bag system for donor blood collection (*see* Note 2).
20. Safety equipment: cryogloves, goggles, and so on.

2.3. HES Method for 1–5-mL Samples of Whole Blood—ABSD

1. Cryovials of varying sizes from 2-mL capacity upwards (Nunc [Roskilde, Denmark] or Costar [High Wycombe, UK]).
2. HES 40% (w/v) solution in isotonic saline (Laevosan).
3. A wire cage with lid to allow total immersion of cryovials.
4. A wide-necked liquid nitrogen container (Jencons, Leighton Buzzard, Beds, UK).
5. A circulating water bath set at 43.5°C.
6. Hand tongs.
7. Protective clothing: cryogloves and goggles or visor.

3. Methods

3.1. RBC Cryopreservation Using Low Concentration Glycerol—ABSD

1. Preparation of RBCs to be frozen: Weigh all the units of blood. If the unit weighs 320 g or more proceed as follows.
2. Place the unit of blood in an individual plastic protective bag.
3. Load the unit of blood into a plastic centrifuge liner. Balance each pair of liners.
4. Place balanced centrifuge liners opposite one another in the Beckman J6 centrifuge.
5. Set the centrifuge to a temperature of 4°C, a speed of $3000g$, and a duration of 15 min.
6. After centrifugation, remove the centrifuge plastic liners taking great care not to disturb the contents. Then remove each blood pack and place it in an extended Fenwal Plasma Extractor.
7. Clamp the end of the pack tube with a pair of forceps and then attach a transfer pack.
8. Release the forceps and express the plasma out of the blood pack into the transfer pack.
9. When all the plasma has been expressed, clamp the end of the pack tube with forceps. Detach the transfer pack.
10. Remove the blood pack from the extractor. Using a heat sealer, seal the end of the pack tube between the forceps and the blood pack. Remove the forceps from the pack tube.
11. Weigh the packed cells and collection bag and record the gross weight on the bag. If the unit weighs less than 320 g, simply record the weight on the bag.
12. Procedure for cryoprotection of packed cells: These procedures are to be performed under a laminar flow hood in a clean room. Disposable gloves and a face mask are to be worn throughout the procedure. Transfer all the equipment/materials into the clean room and place them on a bench opposite the laminar flow hood.
13. Enter the donation number of each unit of packed cells into the Frozen Blood Register. Include the batch number and expiration date of the cryoprotectant, the Air Inlet Sets Travenol 30413, the Air Filters Acrodisc 0.2-µm filter, the Gambro DF1200 Bags, Plasma Transfer Sets Baxter C2243.
14. Fit a bottle hanger to a 500-mL bottle of cryoprotectant and place it under the laminar flow hood.
15. Insert the airway needle through the rubber bung of the cryoprotectant bottle. Fit the 0.2-µm bacterial filter to airway set and ensure that the airway set is closed.

16. After ensuring that the roller valve is closed, insert the transfer set plastic coupler through the bung of the cryoprotectant bottle.
17. Open the plastic seal of one of the blood pack ports and insert the plastic coupler of the transfer set into the port of the blood pack to be cryopreserved. Place the blood pack on a rocker.
18. Invert the cryoprotectant bottle and hang it on a hook under the laminar flow hood.
19. Switch the rocker on.
20. Release the roller clamp and open the airway set on the transfer tube. Allow 250 mL of the cryoprotectant to drop into the packed cells about 8 mL/min. The procedure should take 30 min to complete. After the addition of 250 mL of cryoprotectant, close the roller clamp on the transfer tube.
21. Transferring cryoprotected packed cells into the freezing-bag: Label the Gambro DF1200 bag with a waterproof ball point with the donation number, the ABO/Rh blood groups, the date of donation, and the date of freezing.
22. Remove the transfer plastic coupler from the cryoprotectant bottle and insert it into the left-hand port of the Gambro Bag, having first removed the plastic cover.
23. Transfer the cryoprotected cells from the blood pack into the Gambro Bag. On completion of the transfer, squeeze any air from the Gambro Bag back into the donor pack and quickly close the roller clamp.
24. Seal off the inlet port of the Gambro Bag with the Gambro Sealer. Seal the port again adjacent to the initial seal, and cut off the waste section (*see* Note 3).
25. Weigh the empty donor pack and subtract this from the gross weight of the donor pack, thereby deriving the net weight of cells to be frozen. Record this weight in the Frozen Blood Register.
26. Lift the open edge of the freezing frame and insert a Gambro Bag containing cryoprotected red cells. Ensure that the bag rests as flat as possible within the frame, with the ports sticking out of the top.
27. Clip the open edge of the freezing frame with two small bulldog clips.
28. Immerse the freezing frame in the liquid nitrogen in the LNR 40 with the port end uppermost. Leave the frame static in the liquid nitrogen for 5 min until all the bubbling has ceased (*see* Note 4).
29. Remove the freezing frame from the LNR 40. Then remove the frozen Gambro Bag from the freezing frame and place in the vapor phase of an LNR within 30 s of removal from liquid nitrogen for long-term storage.
30. Thawing of cells prior to washing: Remove the first two units to be thawed from the LNR. Normally, the first and second units are thawed together, the third and fourth after the first two units have been com-

Cryopreservation of Human Red Blood Cells 241

pleted, then the fifth and and subsequent units when the third and fourth units have been completed.

31. Immerse the units in the water bath for 15 min or until completely thawed. Remove and blot off excess water with a tissue.
32. Fill a bath with sufficient alcohol/water/hibitane mixture to cover the pack completely. Immerse each pack one at a time, for 3 min. Remove and place on the metal lid of the bath (*see* Note 5).
33. Check the documentation and sign and date the relevant columns of the Frozen Blood Register.
34. Washing thawed red cell concentrate using the COBE: A disposable face mask and gloves are to be worn throughout the procedure. The hydraulic system of the COBE 2991 cell washer must be primed using the following procedure. This protocol must be carried out at the beginning of each batch of cells to be washed, but there is no need to reprime the machine between individual red cell unit washes on the same day.
35. Turn on the power and set the centrifuge speed to 3000 rpm, super out rate to 450, super out volume to 500, and the pump restore rate to 0. Place auto/manual switch to AUTO.
36. Open the sliding cover to reveal the centrifuge bowl, and remove the lid and alignment blocks. Place the priming cushion in the bowl with the priming disc on top and replace the lid. Close the sliding covers. Press stop/reset followed by start/spin, allow the centrifuge to spin for 20 s and then press stop/reset. Repeat this section at least twice. If the centrifuge bowl has been removed, it may require five complete prime cycles.
37. To check that the priming procedure has been successful, with the priming cushion and disc still in place, press start/spin and allow the centrifuge to spin for at least 15 s. Press super out. If the excess pressure light comes on within 15 s, then the machine has been successfully primed, if not, the machine is not primed and the priming procedure must be repeated.
38. Once the priming procedure is completed successfully, press stop/reset. Allow the centrifuge to come to a halt and remove the priming cushion and disc. Reset the Pump restore rate to 450.
39. Unpack the processing set and ensure that no seals have been damaged. If any are broken then discard the set. Label the hexagonal rotating seal with the number of the unit to be processed.
40. Using Spencer Wells forceps, or the clip provided, close off the blue-striped tube as close to the five-way junction as possible.
41. Press stop/reset, followed by tube/load.
42. Position the five-way junction block slightly above the red cell detector and push the tube connected to the "doughnut" bag into the slot in the detector.

43. Set the valve selector at V1 and insert the red striped tube into the red coded pinch valve, V1. The purple striped tube is then placed into the purple-coded pinch valve. Red indicates the tube connected to the blood bag, and purple is the tube attached to the supernatant collection bag.
44. With the valve selector in the V2 position, insert the green-striped tube in the green-coded pinch valve (V2).
45. With the valve selector switch in the V3 position, insert the yellow-striped tube in the yellow-coded pinch valve (V3).
46. Press stop/reset. This will close all the valves just loaded.
47. Mount the supernatant collection bag on the left-hand side of the machine by fitting the three holes in the top of the bag over the three studs.
48. Open the sliding covers on the machine and remove the centrifuge cover by lifting it with a counterclockwise twist, at the same time lifting the locking catch.
49. Mount the centrifuge cover in the holder at the base of the console. Remove the two white alignment blocks from the centrifuge bowl.
50. Roll the "doughnut" bag into a tight cone around the hexagonal rotating seal, and pass it through the hole of the centrifuge cover. To make this easier, the cover should be slid part way out of the cover holder, or it may be held in the hand.
51. Install the blood-processing bag in the centrifuge bowl by positioning the four holes in the bag over the four studs on the centrifuge. Make sure the bag lies flat over the base of the bowl by pressing the outer edge of the processing bag completely into the bowl.
52. Position the two white alignment blocks around the inlet stem of the processing bag with the widest channel downward around the rotating seal.
53. Replace the centrifuge cover over the alignment blocks and the four positioning studs, and lock in place with a clockwise twist.
54. Ensure that the seal weight slides freely on its vertical guides before fitting the hexagonal seal into its holder by sliding the rear cover forward and lifting the weight. Close the front sliding cover by pushing up to the rear cover. The ready light should now be on. If not, open the front cover and reclose it.
55. Once the processing set is installed and has been checked thoroughly for twists and kinks, the wash solutions and blood can be connected and loaded.
56. Cut or tear one of the plastic closures around the entry ports of wash solution A (hypertonic saline). Remove the cap from the spike on the green tube and pierce the seal of the wash solution. Invert and hang above the right-hand side of the machine, from one of the hooks provided.
57. Snap off the inlet cover from the container of wash solution B (isotonic saline). Remove the cap from the spike on the yellow-striped tube and

Cryopreservation of Human Red Blood Cells 243

insert into the wash bottle. Invert and hang next to wash solution A. Insert a sterile air inlet into the uppermost surface of the container.

58. Remove the cap from the spike of the red-striped tube and insert it into the seal of the PALL SQ40S blood filter. Insert the spike on the other end of the filter into the outer port of the Gambro bag containing the thawed blood. Invert the bag and hang above the left-hand side of the machine.

59. Press blood in. Valve V1 opens and allows the blood to flow into the bag mounted in the centrifuge bowl. Once the blood stops flowing into the centrifuge (even though there may be some blood remaining in the Gambro bag) press air out. This will start the centrifuge spinning and force air out of the processing bag. When all the air is removed, press blood in and any blood forced out of the processing bag, together with any left in the Gambro bag, will flow into the "doughnut." Do not press stop/reset when still in air out mode. Should a large amount of blood remain in the Gambro bag, this procedure can be repeated, but inevitably a small amount will remain in the Gambro bag.

60. Once all the blood has flowed into the "doughnut," press stop/reset. Change the valve selector to the V2 position and press tube load to allow wash solution A to run into the processing bag. Open the covers and mix the bag contents by gently rotating the centrifuge by hand. Once the hypertonic saline has stopped flowing, close the covers and press stop/reset. Return valve selector to V1, and press stop/reset.

61. The machine is now ready to begin washing. The COBE is normally used in the automatic mode. In this mode the functions are controlled by the setting of pins in the program select board. The normal settings used are shown in the manufacturer's instruction manual. The settings should be checked before automatic processing begins.

62. Press start/spin. The COBE 2991 now proceeds automatically. The blood is centrifuged, washed, and the supernatant fluid expressed into the collection bag after each wash. The washing procedure will require more than one container of isotonic saline (wash solution B), therefore as each container is emptied, press hold, replace empty container with a fresh one (including the air inlet), and then press continue.

63. The supernatant fluid from the last wash is to be examined visually as it passes through the purple-striped tube to the waste bag. Should there be excess hemolysis of the cells (i.e., the color intensity of the fluid is high), the matter should be reported to the processing manager. The procedure is completed with packed cells in normal saline with a packed cell volume (PCV) of about 75–85%.

64. Quality control procedure: Measure the volume of washings from the COBE machine, and record the volume. Take a 10-mL sample of the

washings and put it into a universal container. Dilute 2 mL of the washing sample with 8 mL of diluent reagent, mix, and stand for 2 min. Read the optical density of the sample in a colorimeter at 540 nm, using high and low standard hemoglobin solutions to construct a standard curve. Calculate the total quantity of hemoglobin in the washings. If the total is more than 10 g, the pack should not be issued for transfusion.

3.2. HES Method for Whole Donations—ABSD

Unlike the previous method, HES is an external, as opposed to an internal, cryoprotectant and does not require to be removed from the RBCs prior to transfusion. The method is static in that it involves no machinery for freezing or thawing. The procedure is covered by UK Patent PCT/GB 90/0140 filed on February 8, 1989 (*see* Notes 1 and 2).

1. Preparation of the filtered red cell/HES mixture: Fresh whole donor blood units, preferably less than 24 h old, are suspended upside down 2 m above ground level.
2. Under sterile conditions dock the bag to the white-cell filter system and allow the whole blood to pass through the filter into the filtrate bag under the effect of gravity (*see* Note 6).
3. Seal the connecting tube between the filter and the filtrate bag, cut off and discard the filter and original collection bag.
4. Place the filtrate bag in the centrifuge and spin at 4000 rpm for 20 min (*see* Note 2).
5. Express the plasma from the filtrate bag into the satellite bag. It is important that all visible plasma is expelled from the filtrate bag to ensure that a PCV of at least 90% is achieved. Detach the plasma bag and send it for separate processing (*see* Notes 2 and 7).
6. Connect the filtrate bag to the Nuplastic tubing of the freezing bag. Invert the filtrate bag and suspend it 2 m above ground level. Allow the packed red cells to flow into the freezing bag.
7. Express any air that has collected in the freezing bag back up the tubing into the filtrate bag together with a small volume of red cells.
8. Aseptically connect the tubing of the HES pouch to the Nuplastic tubing of the freezing bag. Suspend the inverted pouch 2 m above ground level and allow the HES solution to flow into the freezing bag, displacing all the red cells from the Nuplastic tubing into the freezing bag. Roll up the pouch gently to ensure that the pouch is emptied.
9. Seal the Nuplastic tubing with the Sebra sealer, leaving as great a length of tubing as possible attached to the freezing bag.

Cryopreservation of Human Red Blood Cells

10. Place the appropriate ABO/Rh group label, donation number label, and product expiration date label on the pack.
11. Mix the contents of the freezing bag thoroughly but carefully by repeated manual rotation and inversion of the bag for 4 min.
12. Freezing the HES/red cell mixture: Place the freezing bag into the base of the freezing frame. Ensure that the bag is laying smoothly and centrally and that the tubing/insert area is positioned within the frame recess.
13. Place the top of the freezing frame in position and secure this with six metal clips placed round the edges of the frame.
14. Carefully drop the frame vertically into the LNR40 so as to totally immerse the frame. The nitrogen will boil vigorously. Bubbling should cease in approx 30 s. Leave the frame totally immersed for another 30 s.
15. Using hand tongs, lift the frame from the LNR40. Quickly place the frame on the steel table, unclip the frame, and remove the freezing bag. Transfer the freezing bag to the appropriate storage container. This stage is to be completed within 30 s to avoid the risk of premature thawing.
16. Storage of HES cryopreserved red cells: Bags are to be stored flat and not on edge. Alternatively they may be suspended by the holes in the corners of the bag surround.
17. If storage is in an LNR, the bags must be stored in vapor phase, not actually in the liquid nitrogen (*see* Note 8).
18. Preparation and storage of reference samples: The preparation, freezing, and thawing procedures for reference samples are as described in Section 3.3., steps 1–9, with the addition of the following preliminary steps: Take one cryovial of minimum size 2.5 mL. Cut the tube of the discarded HES pouch and allow at least 0.5 mL of starch solution to run into the cryovial. Take the discarded filtrate bag, cut the end off the tube, and allow a volume of red cells, twice the amount of HES, to run into the cryovial. Ensure that the label details on the cryovial correspond with those in the main freezing bag. Store the vials in the vapor phase of liquid nitrogen in an appropriate racking system.
19. Thawing HES cryopreserved red cells: Check that the temperature of the water in the circulating water bath is within 1° of 43.5°C.
20. Remove the required frozen unit of cryopreserved red cells from the LNR or mechanical freezer.
21. Place the unit in the water bath within 30 s of removing it from the refrigerator, ensuring that the unit is totally immersed.
22. Holding the bag horizontally, move it gently back and forth during the first minute of thawing (*see* Note 9).

23. Leave the bag in the water bath for a further 9 min to allow the temperature to equilibrate and then remove the thawed unit from the bath. Blot it dry with a paper towel. The unit is now ready for quality assessment.
24. The freezing bag is then connected to a transfer pack using a sterile connecting device. The HES/RBC mixture is then run into the transfer pack. This bag may contain plasma or an optimal additive solution.
25. Quality control tests for HES cryopreserved red cells.
    a. Saline stability test: Once the HES/RBC mixture has been transferred into the transfer pack, retain the empty freezing bag and run residual RBCs from the Nuplastic tubing into a universal container. Place 25 mL of isotonic saline into another universal container. Add 500 µL of cryopreserved red cells to the isotonic saline and mix gently but thoroughly. Transfer 10 mL of the red cell/saline mixture into each of two test tubes labeled "1/2" and "2" for the .5-h and 2-h tests. Transfer 2 mL of the remaining RBC/saline mixture to a third test tube. Make up to 10 mL with distilled water, mix well, and label the tube "total." After 30 min, centrifuge the test tube labeled "1/2" at 3000 rpm for 10 min. Transfer the supernatant from the tube "1/2" to the colorimeter and read the optical density (OD) at 540 nm. Likewise read the OD of tube labeled "total." After 2 h, centrifuge the tube labeled "2" and read the OD as for tube "1/2." From these readings calculate the .5-h and 2-h saline stabilities of the red cells as follows (*see* Notes 9 and 10):

    $$(\text{Supernatant OD} \times 100/\text{Total hemoglobin OD}) \times 0.2 = \% \text{ lysed cells}$$

    $$\text{Saline stability } \% = 100 - \% \text{ lysed cells}$$

    b. Plasma stability test: Use step 25a, but substitute isotonic saline with red cell free nonlipemic plasma. In addition, the blank should consist of plasma only (*see* Note 11).

3.3. HES Method for 1–5-mL Samples of Whole Blood—ABSD (see Notes 1 and 2)

1. Take the unit of blood that is to be aliquoted for reference samples. The unit should be preferably less than 24 h old.
2. Take the requisite number of cryovials and to each add the desired volume of blood and half the volume of the 40% (w/v) HES solution.
3. Screw on the cryovial cap firmly and mix by inversion, gently but thoroughly, for 1 min. Ensure that the cryovial is not overfull as this inhibits mixing and does not allow for expansion on freezing.
4. Place the prepared cryovials in a wire cage and close the lid securely.
5. Immerse the cage in liquid nitrogen so that the cage is totally immersed and the tubes are not floating above the surface.

6. Once all the bubbling has ceased, remove the cage using the hand tongs. Transfer the cyrovials to an LNR within 30 s. Store in the vapor, not the liquid, phase.
7. To thaw: Remove the required number of cryovials from the LNR, put into a wire cage, and rapidly immerse the cage fully in a circulating water bath at 43.5°C. Leave the cage in position for 10 min (*see* Note 9).
8. Approximately 80% of the red cells will survive this procedure (*see* Notes 12 and 13).
9. To remove most of the hemolyzed RBCs and HES, rinse the mixture twice in isotonic saline. Alternatively, dilute the mixture directly (1:20) with isotonic saline, which will produce a cell suspension suitable for blood grouping.

4. Notes

1. The HES method has a number of advantages over the presently available methods using glycerol. These include:
   a. Glycerol is a substance which is toxic to the body and therefore has to be washed out of the cryopreserved RBCs before they can be transfused. This technique requires sophisticated equipment and the procedure is most safely performed in a clean room suite. This means that the method is not easily adapted to field conditions. As HES is a substance resembling liver glycogen, it is nonimmunogenic and does not require removal prior to transfusion. The amount of equipment required is minimal, and packs can even be safely thawed in a bucket or tub of hand hot water with virtually no detectable deterioration of in vitro quality control results. This makes the method ideal for disaster or war situations.
   b. The degree of technical skill required to operate the equipment associated with the glycerol method is high, whereas personnel, with no prior laboratory experience, can be fully trained in the HES method in less than 2 d. Two skilled technicians, using four washing machines, can produce 50 U of deglycerolized RBCs in a 12-h shift. In contrast, two relatively unskilled operatives, each with two simple water baths, can produce 600 U in a similar time period using HES frozen blood.
   c. Deglycerolized RBCs are suspended in saline, whereas the ABSD method produces RBCs in a solution of HES, an artificial plasma expander. As HES is used in a 40% (w/v) solution, water is drawn in from the extravascular space to reduce the tonicity of the mixture. This means that a unit of HES cryopreserved RBCs has a volume expansion potential, virtually equal to that of whole blood. The yield from a typical glycerol pack is around 82% of the starting product. With HES packs, only 5 or 6% of cells will not be stable after infusion and approx 90% will be in the circulation 24 h after transfusion.

d. As the washing and deglycerolization of RBCs is technically an open method, the postthaw shelf-life is 24 h. This markedly limits the use of such a process in a disaster situation, as the thawing has to be performed very close to the surgical facility, or the transport to the surgical facility has to be very rapid. The HES technique maintains the microbiological integrity of the system and so the shelf-life is not shortened owing to dangers of bacterial contamination.

e. The volume of each HES/blood pack is approx two thirds of that of a glycerol pack when freezing the same amount of red cells, so storage space and costs are proportionately reduced.

2. Cryopreservation of animal blood: Modification of the protocol is required for cryopreservation of sheep red cells as follows:

 a. CPD or CPDA cause slight damage to sheep cells so heparin is the anticoagulant of choice, but it is vital that special care is taken when mixing during donation.

 b. Sheep cells are spherical and only about one-third of the volume of human red cells, so the plasma is more difficult to separate. It is therefore necessary to centrifuge for about 25 min to pack the sheep red cells to 90% PCV.

 c. For cryopreservation of pig cells, which are a similar size to human red cells, the main requirement is to remove all the plasma by saline rinsing prior to the addition of HES. If all the plasma is not removed, progressive hemolysis occurs, possibly owing to complement activation.

3. Before disposing of the empty donor pack, crosscheck the details on the donor pack with those on the Gambro Bag and the corresponding entry in the frozen blood register.

4. Cryoprotected red cells are to be frozen within 2 h of the separation procedure.

5. When all the units in the batch have been thawed, the water bath is to be emptied and cleaned with hibitane and Meths solution.

6. Whole blood filtration: Prior to cryopreservation it is essential to remove white cells and platelets. The contents of the white cells and platelets are highly thromboplastic, and as these cells are destroyed on freezing, thereby liberating their contents into the HES/RBC mixture, failure to filter could cause a disseminated intravascular coagulopathy when the RBCs are subsequently thawed and transfused.

7. Interaction with plasma proteins: The removal of the bulk of the plasma from the red cell concentrate for separate processing means that interaction of HES with some plasma proteins is reduced to a minimum. It has been found that where the concentration of HES in the solution is less than 14%, there is no interaction with plasma proteins, but there is a reversible

precipitation of some proteins with the 40% (w/v) HES solution *(6)*. In the ABSD method, the overall percentage of HES in the frozen/thawed packs is only 6% and the starch is rapidly diluted further in the bloodstream of the recipient.
8. Storage: Both human and sheep red cells have been stored in liquid nitrogen vapor phase without deterioration for up to 12 yr, as demonstrated by in vitro testing. In theory, periods of storage for stockpiling of rare groups could be as long as 2000 yr. Storage in mechanical freezers has been carried out at ABSD at the following temperatures without deterioration of the red cells as assessed by in vitro quality control tests, for the undermentioned periods: $-140°C$ for 4 mo, $-120°C$ for 4 mo, $-100°C$ for 4 mo, and $-90°C$ for 5 wk. At storage temperatures of -80 and $-85°C$, progressive deterioration occurred within a matter of days. Further trials covering the range -90 to $-120°C$ are currently being undertaken at ABSD.
9. As the product is not washed after thawing, only a very small amount of hemolysis can occur before the resultant product would be unsuitable for transfusion. To achieve recoveries in excess of 98%, rapid freezing is essential.
10. Postthaw storage: After thawing, the pack of HES/blood can be stored at 4°C for 10 d without deterioration. Additional tests may be performed to assess ATP and 2,3 diphosphoglycerate (DPG) levels *(7)*. P50 levels have also been measured at ABSD. These parameters have been found to be only slightly affected. If mixed with a diluent or optimal additive solution storage times at 4°C are not improved over those of the mixture stored in the undiluted state.
11. Plasma stabilities: Plasma stabilities may be slightly higher than saline stabilities because plasma factors allow a few very slightly damaged cells to recover. The percentage recovery of intact cells, immediately postthaw, is not used as a quality control procedure, because it is only a crude indicator of pack quality. HES coats the surface of the RBCs and may provide a scaffolding for damaged membranes so that some cells appear intact, although they will rupture if diluted with isotonic saline.
12. Cryovial reference sampling: Processing cryovials using the described methodology will give a postthaw yield of around 80% saline stable RBCs. This yield is lower than that in the corresponding freezing bag because of the slower rates of freeze and thaw, and because any air in the tube acts as an insulator. In addition the film may be thicker than that in the bag. In thick films, as the outer layers of the mixture freeze first, an inner liquid core is left. As this freezes there is no room for expansion and crush damage will occur. It is possible to refreeze the cryovial, which on subsequent thawing should produce around 60% saline stable cells and is still an ample yield for crossmatching.

13. When processing cryovials, a visor should be worn in case any liquid nitrogen penetrates the cap area of the cryovial. On thawing, rapid expansion of the gas could cause the occasional cryovial to rupture.

References

1. Wise, R. (ed.) (1970) *Frozen Blood: A Review of the Literature.* Gordon and Breach, Reading, PA, pp. 1949–1968.
2. Mishler, J. M. (1982) *Pharmacology of Hydroxyethyl Starch,* Oxford University Press, Oxford, UK.
3. Meryman, H. T. and Hornblower, M. A. (1972) A method for freezing and washing RBCs using a high glycerol concentration. *Transfusion* **12,** 145–156.
4. Valeri, C. R. (1982) *SOP Red Blood Cell Collected in the CPD-A 800 mL Primary PVC Bag Collection System and Stored for 3–35 Days.* Naval Blood Research Laboratory, Boston University School of Medicine, Boston, MA.
5. Thomas, M. J. G. (1990) Military experience in blood supply applicable to disaster medicine, in *Transfusion in Europe* (Castelli, D., Genetet, B., Habibi, B., and Nydegger, U., eds.), Paris: Arnette, pp. 237,238.
6. Bell, S. H. (1985) *The Interaction of Hydroxyethyl Starch with Plasma Proteins.* Masters thesis submitted to Brunel University, Uxbridge, Middlesex, UK.
7. Oglesby, D. (1993) *The Assessment of Frozen/Thawed Hydroxyethyl Starch Cryoprotected Red Cells When Stored at 4°C for 21 Days.* Masters thesis, University of Portsmouth, Portsmouth, UK.

Index

A
Algae,
  growth phase, 86
  preculture, 85
Alginate beads, 127
Ampules,
  labeling, 35
Animal blood cryopreservation, 248
Antigenic activity,
  preservation of, 18
Avian spermatozoa,
  artificial insemination 173, 175
  diluent, 170, 174
  semen collection, 169, 170
  thawing, 172, 173

B
Bacteria,
  cross-contamination, 22
  storage period, 22
Budapest Treaty, 31

C
Centrifugation,
  human erythrocytes, 239, 241
  human spermatozoa, 231
  mammalian cell lines, 185
  plant cell suspensions, 107, 115
  thylakoids, 77
  viruses, 19
Cold hardening,
  fungi, 57
  protoplasts, 98
  thylakoids, 78
Control of Substances Hazardous to Health (COSHH), 17, 56

Cooling chamber preparation, 148
Cooling rates,
  algae, 87
  amoebae, 68
  avian spermatozoa, 172
  bacteria, 29, 30
  fish spematozoa, 159
  fungi, 50
  human spermatozoa, 228
  livestock semen, 194, 195
  mammalian cell lines, 184
  mammalian embryos, 203, 204
  mammalian embryos vitrification, 214, 218
  oyster spermatozoa, 147
  plant cell suspensions, 106, 118
  plant seeds, 139, 142
  protoplasts, 99
  yeasts, 40
Cryoadditives, see Cryoprotectants
Cryomicroscopy, 99
Cryopreservation,
  general principles, 92–94
Cryopreservatives, see Cryoprotectants
Cryoprotectants,
  algae, 83
  amoebae, 67
  avian spermatozoa, 170, 174
    contraceptive action, 167
  fish spermatozoa, 152, 161, 162
    equilibration time, 162
  fungi, 50, 58
  livestock semen, 193
  human embryos, 224
  human erythrocytes, 237, 238

251

mammalian cell lines, 180
mammalian embryos,
   glycerol, 203, 204
   propanediol, 204, 205
oyster spermatozoa, 146
plant seeds, 137
plant shoot-tips and meristems, 123
protoplasts, 94, 95
thylakoids, 74
yeasts, 45
Cryostorage, *see* Cryopreservation
Cryotubes, *see* Ampules
Cryovials, *see* Ampules
Cysts,
  amoebae, 66

D
Documentation, *see* Record keeping
Dormant plant buds, 121

E
Equipment sterilization, 24
Erythrocytes, *see* Human erythrocytes

F
Fish spermatozoa,
  extenders, 151 155
  field freezing, 152
  gamete procurement, 154, 155
  warming/thawing, 159
  prefreeze quality, 160, 161
Freeze-dryer,
  centrifugal, 22
  shelf, 22, 55, 56
Freeze-drying,
  algae, 81
  general principles, 21–23
  reconstitution,
    fungi, 55
  spin, 54
  secondary drying, 54
  viruses, 15–17

G
Genetic stability,
  amoebae, 68
  bacteria, 29
Glass beads,
  use in storage of bacteria, 25, 29

H
Human embryos,
  blastocyst stage, 226
  cryovials, 225
  culture media, 223
  early cleavage stage, 226
  equilibration, 231
  straws, 225, 226
Human erythrocytes,
  glycerol,
    low concentration, 237, 238
    toxicity, 247
  hydroxyethyl starch (HES), 238, 244–247
  plasma stabilities, 249
  thawing, 240, 241, 245
  washing, 247
Human oocyte, 222
Human spermatozoa,
  buffer, 224
  collection and analysis, 228
  postthaw insemination, 228, 229
  sublethal cryodamage, 221
  thawing, 228

I
Ice crystallization, *see* Ice nucleation
Ice nucleation, 108, 109, 194, 203, 206, 207
Ice seeding, *see* Ice nucleation
In vitro fertilization (IVF), 231

L
Liquid nitrogen,
  safety, 45, 58

Index

Livestock semen,
  antimicrobial agents, 193
  buffers, 191
  cooling, 191, 194
  diluents, 192
  energy substrates, 192, 193
  equilibration, 191, 194
  postthaw fertility, 194
Lyophilization, *see* Freeze-drying

M

Mammalian cell lines,
  avoidance of antibiotics, 185
  cross contamination, 186
  culture density, 186
  growth phase, 180
  handling, 180
  microscopic examination, 181
  screening for mycoplasma, 186
  serum, 185
  thawing, 183
  vital staining, 181
Mammalian embryos,
  blastocyst vitrification, 216
  1-cell embryo vitrification, 216
  2-cell embryo vitrification, 216
  4-cell embryo vitrification, 216
  8-cell mouse embryos, 202
  cryoprotectant removal, 204, 218
  developmental stage, 205, 206
  embryo banking, 205
  oocyte vitrification, 216
  postthaw recovery, 215
  prefreeze quality, 205
  straw loading, 206, 214
  thawing, 214
  vitrification cryotubes, 217
  vitrification solutions, 212, 213
    exposure duration, 216, 217
Moisture content,
  bacterial cultures, 29
  fungi, 59
  plant seeds, 134, 135, 137, 141

N

Nucleic acid,
  preservation, 18
Nucleation, *see* Ice nucleation

O

Oyster spermatozoa,
  adult conditioning, 146
  gamete extraction, 146
  thawing, 147

P

Plant axillary meristem, 121
Plant cell suspensions,
  preculture, 106, 108, 116
  standard freezing protocol, 104
    improvement, 109
  vitrification solutions, 116
Plant meristem,
  dissection, 123
  encapsulation, 122, 124, 127
Plant seeds,
  longevity under liquid nitrogen, 133
  rehydration, 138, 142
  surface sterilization, 138
Plant shoot-tips and meristems,
  cryoprotective dehydration, 125
  preculture, 123
  vitrification, 123, 124, 126, 127

R

Record keeping, 18, 29, 46, 148, 160, 182, 249

S

Safety,
  amoebae, 66
Serodiagnostic preservation, 18
Sporulating fungi, 57
Storage temperature,
  algae, 87
  bacteria, 29

fungi, 59
human erythrocytes, 249
livestock semen, 192
mammalian embryos, 203
oyster spermatozoa, 147
plant cell suspensions, 106, 109
plant seeds, 139
viruses, 19
yeasts, 40, 46
Straws,
  polypropylene, 41
  sealing, 41, 115, 214
  thawing, 42
Survival *see* Viability
Suspending fluid,
  bacteria,
    freeze-drying, 24
    cyropreservation, 24
  yeast,
    freeze-drying, 32

T
Thawing rate,
  algae, 85
  fungi, 51
  livestock semen, 195
  mammalian embryos, 203
  plant cell suspensions, 109
Thylakoid,
  isolation, 75
  membrane integrity, 76
    cyclic photophosphorylation, 76
    plastocyanin release, 77
    volume measurements, 76, 77
Thylakoids,
  anion concentration, 78
  chlorophyll concentration, 78
  magnesium concentration, 78
Trophozoites, 63

V
Vacuum testing, 27
Viability,
  cutoff values for collections, 3, 81

Viability assessment,
  algae,
    vital staining, 86
    colony formation, 86
  amoebae, 68
  avian spermatozoa, 173, 175
    insemination, 160
    motility, 160
  fungi, 51
  human embryos, 230
  human erythrocytes,
    plasma stability, 246
    saline stability, 246
  mammalian cell lines, 181, 186
  mammalian embryos,
    implantation, 204, 215
    in vitro culture, 204, 215
  oyster spermatozoa, 149
  plant cell suspensions, 107, 110
  plant seeds,
    vital staining, 137
    germination, 138
  plant shoot-tips and meristems,
    127, 128, 131
  protoplasts, 96, 98
  yeasts, 44
Virus,
  arbo, 8
  smallpox, 8
Viruses,
  general handling, 11
  infectivity titer, 19
  safety, 8
  storage periods, 19
Virus storage,
  optimal pH, 18
Vitrification,
  general principles, 93, 94, 211, 212
  plant cell suspensions, 113–115
  protoplasts, 96, 97

Y
Yeast,
  rehydration, 34